Studies in Post Colonialism

Studies in Post Colonialism

Lilack Biswas

PARTRIDGE
A Penguin Random House Company

To order additional copies of this book, contact
Partridge India
000 800 10062 62
orders.india@partridgepublishing.com

www.partridgepublishing.com/india

Dedicated to the Lotus Feet of Mahatma Shri Narayan Goswami
for being the Friend Philosopher and Guide and showering
his limitless bliss on such an inferior soul like me.

I express my cordial thanks to Professor Dr. Jayanta Mete
for his continuous inspiration and encouragement

Contents

Introduction

Post colonialism, as defined by the oxford dictionary is "The political or cultural condition of a former colony" and "A theoretical approach in various disciplines that is concerned with the lasting impact of colonization in former colonies". The former definition is highly restricted to the political condition of a former colony, or it may extend its scope to the economic and social scenario as well. In the definition as a theoretical approach, it covers a wider area. If we deduct the phrase "lasting impact of colonization" we get the whole matter which is very aptly encompassed in a nut shell. Here again are a number of issues. Is the lasting impact of colonization a singular number? If not what are the impacts? If yes what is the nature of this impact? Is it like a huge tree with a number of branches; but all branches are held together with a single trunk? If it is so, what are the ramifications of the "lasting impact"? In my opinion the last assumption is most suitable to describe the nature and scope of post colonialism; it is a huge tree with complex and overlapping ramification.

The political colonization of any land does not only subdue the native forces from political or economic power but also restrain the cultural and social traditions and progress, instead imposes the culture and social way of life of the colonizer, which after a long time becomes very much homely to the colonized and they almost come to the verge of their cultural extinction. Their literature, their music, painting, sculpture and all other form of arts and crafts come to a new dimension, either good or bad, under the direct influence of the colonizer.

After independence of such colonies, the society of the new born country is at a loss as they have eagerly abandoned the culture of the colonizer and have unknowingly forgotten their own traditions. Perhaps this condition is expressed in the first definition of the oxford dictionary mentioned above. But there is no stopping of the social force. Then from the remnants of the colonizers and the residue of their own they try to rebuild a new culture. At the same time they try to hold fast to the left over of the colonizer and reincarnate their own. It gives a cultural conflict and also amalgamation. The direct effect of this quest for cultural identity of a post colonial society is best seen in its arts and literature as well as all form of media and communication. One example of this may be presented as the huge number of translation from English to Indian languages in independent India. The Indian society is holding fast to the British arts and cultural paradigms but presenting it to Indian people in their vernacular language; such is the case with adaptations. Adaptations also show intellectual and creative bankruptcy. It happens only when a creative mind looses faith in his cultural tradition to find the subject matter of his or her creation, and goes on to re organise the foreign into a native form.

The more time passes the more complex it has become. In present days post colonialism is not only a hangover of the colonial legacy, but also a reaction against the colonial force. It also consists in the search for originality as a reaction to and rejection of the colonizer's cultural and intellectual impacts.

Therefore, in history, literature, film, mass media, architecture, fine and visual arts and crafts of the formerly colonized society we can see the post colonial notion of one kind or another. In this present volume an effort has been made to pin point the post colonial characteristics in the respective work of art and in the treatment of history as well. The keen observation of eminent scholars have enriched the content of this book very much. I extend my hearty thanks and gratitude to all the contributors.

Lilack Biswas

Chapter 1

Salman Rushdie's *Shalimar the Clown*: A space devoid of borders?

Dr. Bipasha Som

Faculty Associate

Department of English and Modern European Languages.

School of Humanities and Social Sciences.

Gautam Buddha University.

Gautam Buddh Nagar.

Uttar Pradesh

It is not a co-incidence that Salman Rushdie's fictional work *Shalimar the Clown*, delineating psychological crisis resulting from loss of identity and roots, is set in Kashmir, a land piece suitable enough to be marked as a physical and tangible element of derision towards the conjecture of nationalized sense of identity and rootedness. Apparently, the novel tells the story of a ruined love, the story of two lovers in the land of raging violence and hatred. But in this apparently simple tale, the author has infused many problematic issues and questions that reveal themselves in layers. And *Pachigam*, a small peaceful village in the state, exemplifying repressions and exclusions that the postcolonial nation imposes on its periphery, works as an apt canvas for the unfolding of the layered narrative of expedition from innocence to betrayal, from being to becoming, from rootedness to rootlessness.

The book in question has treated, with other themes of wider dimensions, the interconnection of two cultures, Eastern and Western, in its text that depicts migration, cultural hybridization, rootlessness and transnational identity, a theme seems suitable enough in this era of new colonialism namely 'globalization'. But while doing so, the author has not presented the story from a single perspective by holding the mirror up to the reality. Whether it is because for him reality as such does not exist, or, he does not wish to represent reality in one particular way, is a matter of deliberation. What is evident is that while producing a master edition or side of the story, he provides many sub editions as well, just as he delineates people with not more than a 'partial' identity; thereby overwhelming the readers as to how to arrive at the total story or the total meaning of the text. The version he presents is, as he himself says "no more than one version of all the hundreds of millions of possible versions".[1] From the very beginning, it makes the reader apprehensive that this fictional world is going to have no fixed centre. Post colonial philosophy, in its attempt to resist marginalization, looks at centre as an oppressing construct, a narrative, an essential device for 'otherization' and not a fixed and unalterable reality. Set in a typical, though unproblematic space marked by difference and ever changing boundaries, with its various interwoven levels of narratives, the account of the fiction reroutes postcolonial paradigm by bringing in different worlds colliding and exploding in a microcosm as well as expanding in a globalized universe. The novel has plot, characters, setting and theme, though with them all, the text has not taken the form of a simple tale with a definite beginning and end, rather it has taken the form of mosaic.

But, if looked at intensively, the novel *Shalimar the Clown*, after all its "textualization of space", if one may be permitted to borrow the term, reveals itself as a text that ends in somewhat of a different note that one can term, with some limitations though, as uncharacteristic of fiction following a postcolonial-postmodern paradigm. There are multiple voices present in the story each telling the same incident in its own way repeating the theme and challenging the reader because it is questioning the way one reads and interprets a text and demanding the reader to be the co-creator of it. But at the end, it seems that the text that is plagued with a crowd of 'nobodies' seeks to grapple with the essential identity of its characters. The reader can afford to create at least something out of nothing and get a sense of fixity and rootedness.

The basic story of *Shalimar the Clown* is a seemingly uncomplicated story of love, passion, honor, betrayal and revenge. Much of this story is personal and intimate. But there is a bigger picture as well. Rushdie frames this story within the contemporary geopolitical context, the tragedy of Kashmir, the needless tragedy of religious hatred. The novel weaves the story of Boonye and

---

[1]   Salman Rushdie, *Imaginary Homeland, essays and criticism 1981-91* (Great Britain: Penguin Books,1992),p. 10.

Shalimar in and out of the story of modern day Kashmir. Both their life and relationship; and the landscape where it bloomed, are full of love and passion as well as violence and hatred. But the author has not delved deep into this world. Rushdie has deliberately left it vague and confused. Even he has not delved deeply enough into the passion that has given birth to the intense and problematic sense of honor and revenge in the protagonist Shalimar, thereby working as a fountain source of all the tumultuous happenings in his life and in this text. One, at this point, needs to hasten to add that it is again a point of deliberation whether Shamilar can be called the protagonist of this text, a large part of which is engrossed in the unstable wilderness of distressed characters flowing into each others all of whom assume equal prominence in the narrative. Nevertheless, Rushdie has not dealt with the workings of human psyche in depth. The novel is conceived on an abstract plane. The author has not come down to the world with its palpable particulars. The novel blends myth and cold reality. Early in *Shalimar the Clown*, a character states the novel's theme:

Everywhere was now a part of everywhere else, Russia, America, London, Kashmir. Our lives, our stories flowed into each other's, were no longer our own, individual, discrete.[2]

Shalimar the Clown is a story of everywhere. California, India, France, Britain, Pakistan, Algeria, even the Philippines. Salman Rushdie depicts the interconnected human variety of the world. It takes place over many years and features Osama bin Laden, Heinrich Himmler, Rodney King and Lord Lucan too. East meets West, North meets South. Transients, migrants and immigrants converge and mingle into something that is in between, dislocated and suspended in the middle of nowhere. The author has done it with an epic cast of characters and numerous side events embedded in the operative plot, the background of history and culture. But the pieces of that epic story do not always fit together properly and the author does not seem to bother.

The novel's trickery however, comes not from the plot, but from the execution of the plot. Rushdie writes stories that flow into each other. The story repeats itself many times along with the multiple quests of its characters. In each interaction they encounter the same elements but differently each time. The world around them itself alters with them, though the physical environment remains the same. The book begins near the end. The central act

---

[2]   Salman Rushdie, *Shalimar the Clown* (New York: Random House,2005), p.47.

of revenge takes place in the first section. The rest of the book describes what led up to it. After briefly describing Ophul's murder, *Shalimar the Clown* circles back in time, interweaving ancient myth, old story and contemporary incident using flash back technique.

The main story of the novel is only a part of the whole. Everything is connected and every one is a part of everyone else. *Shalimar the clown* is truly a work of the era of globalization. It is a complex tangled tale with multiple interlocking episodes. The novel with its five parts, each narrated from a different view point and different time, does not really lead to a well defined end, at the end of the novel we find its characters all broken and confused individuals standing in the midst of no-where. The novel started with a centre, but very soon it gives way to the disintegrating power of the world. The land of Kashmir in general and their village of Pachigam, in particular, used to subscribe to the doctrine of "Kashmiriyat", Kashmiriness, the belief that at the heart of Kashmiri culture there was a common bond that transcended all other differences. "We are all brothers and sisters here," Shalimar's father, Abdullah, the leader of a band of traveling players, proclaims. "There is no Hindu-Muslim issue." But the most part of the novel demonstrates just how wrong Abdullah is. The communal violence in Kashmir prompts Rushdie's characters to think about the influence of planets on humans and repeating cycles of creation and destruction.

There planets were the grabbers. They were called this because they could seize hold of the earth and bend its destiny to their will. The earth was never of that kind. The earth was the subject. The earth was the grabee.[3]

The novel that is all about conflict crosses the thin line between Fact and Fiction, when it comes to Kashmir. We encounter a parade of the kind of unnaturalness if not improbabilities – a mullah, made entirely out of iron, inveighs against the infidel at one of the camps Shalimar attends.

In the flow of the abstract, sometimes broken up story, the characters are every so often underdeveloped and not distinct enough. Even from the beginning of the novel the principal characters in *Shalimar the clown* find their identity slipping each moment. In the middle of the bewildering anomalies of life they are engrossed in their failed attempts to find the loose ends of their disjointed rootless lives. Almost all the characters are uprooted from their own

---

[3] Ibid, p.55.

soil, and are in the quest of something almost unachievable. They are looking for their own self, their inner realities and trying to get to the bottom of various contradictions. The novelist makes it a point to uphold only the 'contrariness' of their being and has not taken the pain to give his characters a plausible inner life; as they do not have their inner self intact, instead they have bizarre quirks, unusual looks or magical powers, like the figures in a fable.

The main character Shalimar, a gentle and open-minded individual, a tight rope walker, by profession in the Pachigam village is a confused individual. His character is to some extent unrealized and underdeveloped. Little is revealed about the complex feelings he must have gone through to get from a clown to a killer. He, in the course of time is turned into an Islamic fundamentalist terrorist and a blood thirsty criminal. The reason behind his transformation is the shattering away of his world of love and also the idyllic world of his Kashmir, whose fate runs parallel to his and travels from peace, love and loyalty to dirty lust of power and disloyalty. The author has used this personal tragedy to interrogate the political and religious tragedies of the modern world. All the principal characters' lives are intrinsically intertwined with that of Kashmir and they symbolically represent each-other. One of the female protagonists of the story, a brilliant dancer and wife of Shalimar, Boonyi, is deluded by the dream of something unreachable and on a restless caprice she leaves Shalimar for Max and has to pay for that with her life after living in a live-dead situation for days together and lamenting for what she has done. In her exile at first in Delhi and then in the outskirts of Kashmir itself she was continuously waiting for some one to rescue her. She was waiting for Shalimar, whom she loved once and had married, to come and castigate her and win her from Max, but no body came, she is, just like her motherland Kashmir ravished by foreigners and their greed and may be by the lust of power and the lust of something attainable that is there deep within themselves as well. She started recovering once she is back in her land, but some bruises always remain, the marks of abuse that are never to fade. The anti hero of the work, if one is permitted to call him so, Max Ophul's is 'a man of movie-star good looks,' who grew up in 'a family of highly cultured Ashkenazi Jews' in Strasbourg, fought in the anti-Nazi Resistance in order to protect his identity, making his daring escape from Strasbourg in a Bugatti plane. And may be this octogenarian Max tries to reconstruct his own self and identity by having unusual affairs with much younger girls including India's 'hottest box-office star'.

May be the most crucial symbolism comes with another principal character 'India' who even had the name 'Kashmira' initially given by her mother who wanted her to be a part of her own world. But she is robbed of that identity of hers and placed in a far-off land, virtually in the midst of no-where with a western Jew father and an Indian Hindu mother and a terrorist Islamic step father and yet an English Christian keeper mother constituting her haphazard world. She cannot really identify herself with any one. She cannot even recognize her inner voice which comes out in sleep in the form of sometimes loud grunts sometimes soft whimper. She does not know where she comes from, where she belongs. She has always known that her mother has died while giving her birth and her father is the charming ambassador who is not the way a normal father generally is or should be. The uncertainty in her inside, the state of not knowing herself, betrays itself in the form of anger which she channelizes towards the liking for the martial arts like archery. Archery is the sport she loves. It seems, all through the days of her 24 years life she has been waiting for the time when she will be able to reclaim her own identity through that of her mother. And that time comes when her father is killed by his Kahmiri chauffer and driver Shalimar and the reason behind the murder turns out to be not something political but something very personal, and an act of revenge. When she comes to know about her mother, it is like she comes to know what she herself is. She literally tries to look in the mirror and scrutinize herself in order to see the persona of her mother in her and her own identity as well which is in turn shaped by that of her mother. "Kashmir itself may have been annihilated but the seduction of Kashmir by America has produced a bastard child-India Ophuls a.k.a. Kashmira Noman - a hybrid being, who lives in America and who loves her American father, but who is also in the process of discovering who her father really is,what he has done, and who her mother was" (Teverson. 2005.).[4] She found something; it is as if she found a route to her root. and that route eventually takes her back to her root in the literal sense when she visits Kashmir in spite of repeated warning from the officials about the unrest in that

---

4   Andrew,Teverson.*Rushdie's Last Lost Homeland: Kashmir in Shalimar the Clown*.Vol 1. No 1. http://www.litencyc.com/theliterarymagazine/shalimar.php. 1st December 2005. web. 25th September 2007.

region. She sits beside her mother's grave and "something enters into her". She wept profusely and it is as if she comes out of a years-old shell and becomes herself, recognizes her true self, her life, her joys sorrows and even the aim of that life. And that gives rise to the strong wish to take revenge on the killer of her mother which is, eventually, to round up the story of revenge of her father, who has spent all his life waiting for the right moment to come for him when he will be able to take revenge. Finally she does that, the story of revenge ends with revenge taken on the avenger himself, and at least something puts an end to something, the unending chain of vague blurred insignificance is stopped at a point. The two typical factors compounding the crises of identity in the text: physical and mental specification and an uncertain inheritance, do seem to find a bottom-line. We may even hope that some "happy" ending will also come as, now, after visiting Kashmir Kashmira has fallen in love with an Indian and is even thinking of getting married and settling down.

Going by Satya Mohanty's theory of post-positivist realism, if one may, one can, probably, say that identification with social and cultural roots without re-inscription of the rigid binaries, as the process of forming possible meaningful identities, has found its way in the book where cultural identity is shown not to be determined by birth only but to be continuously reconstructed as one works through the meaning of his/her personal and communal histories. This identity gives a greater coherence to a man's everyday life. The conceptual frameworks, found in this text, do seem to have truth value even when they are born out of claims about experience, memory and identity.

Although questions remain regarding the ideologies of identity in a given work, we may say that the story that begins on the street like a postmodern man lost in the globalized jungle of this 'post national' era, if one is allowed to use that term, is spared the fate of a postmodern man of dying on the street. The "man" returns to his roots and finds himself. In fact the root one can find here is stronger than one can imagine because it has come from the urge for self-realization, an urge that has historically played major role in building national consciousness in a postcolonial society. May be the situation is best illustrated in the author's own words where he says:

"The danger with writing this kind of novel…is that you end up writing about nothing and the book has no roots at all. You just skim the surface of everything and don't get under the skin of anything. So you need more roots".[5]

So more roots is what one needs and Rushdie has been successful to achieve that, in his work, at least partially. It is this quest for roots in part of the author that makes a frenzied torrent of ideas, scenes, and observations spill onto every page, leaving the reader exhausted and exasperated. But that finally reaches, or at least seems to reach a destination, a totality, an end.

---

[5]   Derbyshire, Johnathan. Salman Rushdie: Interview http://www.timeout.com/london/books/salman-rushdie-interview. *Apr 6 2006, web, 19th October 2007*

Reference

1. Barry, Peter. 2002. *Beginning Theory: An Introduction to Literary and CulturalStudies*. 2nd edn. Manchester: Manchester University Press.
2. Belsey, Catherine. 1980. *Critical Practice*. London: Methuen.
3. Culler, Jonathan. 1983. *On Deconstruction: Theory and Criticism after*
4. *Structuralism*. London: Methuen.
5. Culler, Jonathan. 1997. *Literary Theory: A Very Short Introduction*. Oxford: Oxford University Press.
6. Eagleton, Terry. 1996. *Literary Theory: An Introduction*. 2nd edn. Oxford: Basil Blackwell.
7. Furniss, Tom and Michael Bath. 1996. *Reading Poetry: An Introduction*. London: Harvester Wheatsheaf.
8. Jefferson, Ann and David Robey. 1986. *Modern Literary Theory: A ComparativeIntroduction*. 2nd edn. London: Batsford.
9. Lentricchia, Frank and Thomas McLaughlin, eds. 1995. *Critical Terms for Literary Study*. 2nd edn. London and Chicago: Chicago University Press.
10. Mohanty, Satya. 1997.*Literary Theory and Claims of History*.Ithaca: Cornell University press.
11. Selden, Raman, Peter Widdowson and Peter Brooker. 1997. *A Reader's Guide to Contemporary Literary Theory*. 4th edn. Hemel Hempstead: Prentice Hall.
12. Randy,Boyagoda. *Does Rushdie Matter? Celebrity is the enemy of the artist*. Volume 011, Issue07. http://www.weeklystandard.com/Utilities/printer_preview.asp?idArticle=6254&R=C8145223 1st October 2005.web. 5th July 2007.
13. Updike, John *Paradises Lost: Rushdie's "Shalimar the Clown"*. http://www.newyorker.com/archive/2005/09/05/050905crbo_books 5th September 2005.web.19th November 2007.
14. Siegel, Lee. *Rushdie's Receding Talent*. 3rd October, 2005 edition of The Nation. http://www.thenation.com/article/rushdies-receding-talent 15th September 2005. web.7th august 2007Chaudhuri, Amit *Oh, For The Return Of The Clown Prince*.
15. http://www.outlookindia.com/article.aspx?228543 12th September 2005. web. 22nd August 2007.

Chapter 2

Homeless in the Homeland: A postcolonial study of Arundhati Roy's *The God of Small Things*

Bula Rani Howlader
Assistant Professor
Dhrubachand Halder College
24 parganas,(S), West Bengal, India

Introduction:

Booker Prize winner novel *The God of Small Things* (1997) by the Indian novelist and political activist Arundhati Roy is replete with several post colonial issues. It provides a deep insight into the social, political and cultural condition of a once colonized nation by painting the picture of a family living in Ayemenem, a small town in Kerala. The members of the family have a home but still they can be considered homeless in the figurative sense of the term. The term homeland is used for the Ayemenem house, the city, the state and the nation. According to Mental health literature homelessness can be a cause of psychological trauma, but here the psychological trauma is taken as homelessness. So, the figurative use of the term implies a feeling of 'out of place' (Selden 233) both inside and outside home they are suffering from. This homeless state in the homeland leads the central characters to the psychological conflicts and trauma. They are constantly coming up against the Indian societal power structure, class and caste system and at the same time, dealing with the conflicts of identity and cultural belonging. Both contribute to their actual and metaphorical homeless status.

During the troubled period of about 150 years of British rule in India, Indians had to deal with an alien culture, economy and politics. Even after getting independence in 1947 their cultural, economic and political effects are still felt in the society which attracts the attention of the post colonial critics. *The God of Small Things* corroborates that "Culture is not a fixed origin to which we can make some final and absolute Return" (Hall 226).Thus the novel once again shows the cultural legacies of colonialism and imperialism. The hybrid identity of the people of a post colonial nation creates a kind of ambiguity and complication in the establishment of their own native society free from the culture of the colonizer. The power relationship between the colonizer and the colonized develop a post colonial identity crisis in the post colonial subjects. The conjuncture of this identity crisis, hybrid culture and the traditional Indian society contributes a lot to the characters' actual and metaphorical homeless status.

Significance of the narrative:

The intricate, incoherent and fragmented narrative technique used in *The God of Small Things* is symbolic of the different perspectives of the novel. The narrative of the novel itself works as an important character. The novel does not have any prominent protagonist rather the story is told by a non-participant omniscient narrator. So, it becomes difficult for the readers to understand whose perspective the narrator is presenting but as the readers spend significant portion of the narrative with Rahel around readers may think that the story is told from her perspective. But this strategy of telling the story is "in some way related to its role as a trauma narrative" (Longworth 21). The novel being a post-colonial Gothic hybrid juxtaposes the Indian history and the present post colonial identity of the nation. According to Michelle Giles, "Roy employs the Gothic conventions of dark imagery, the supernatural, the haunted house, the ancestral curse, a threatening atmosphere, doubling and incest to personalize larger cultural horrors of India as experienced by one family in Kerala"(Giles). The story is told in flashbacks and necessarily the setting shifts back and forth between 1969 and 1993. "The flashback, it seems, provides a form of recall that survives at the cost of willed memory or of the very continuity of conscious thought" (Caruth 152).This non-linear narrative with the western Gothic elements like hallucinations, history house (may be taken as mystery house), and dreams give a multidimensional perspective of the novel. Thus the narrative style helps to delve deep into each character's perspective demonstrating the reader his or her personal stands. The Gothic elements like death, decay, uncanniness, powerful love that can be traced in the characters help to reflect on the characters' psychological trauma. But unlike the Gothic novel, this novel uses the strategy of Gothic narrative to show the tension between the subjugated, repressed past and the decolonized unstable post-colonial present. As claimed by many critics the Gothic in this novel is seen as directing and calling into question the issues of colonialism and post colonialism. Roy intentionally uses different layers of narratives to keep the dilemma, ambiguity and psychological conflict of the characters and their circumstances. The narrative style and structure thus carry the interrupting symptoms of trauma and "the childlike perceptions represent the nature of pre-narrative integrated traumatic memories (Longworth 21)".

The family and the social structure:

Let us discuss how the social and political forces in a post-colonial nation affect the lives of the individuals. The novel is basically a story about an upper-caste Syrian Christian Ipe family in a small town, Ayemenem situated in the state Kerala. Selection of this state located at the south-western part of India is also meaningful from the perspective of our study. Firstly; a bit of biographical criticism can be done here. Arundhati Roy herself was born to a Bengali Hindu father and Keralite Syrian Christian mother. Naturally she was familiar with the Syrian Christian ideals and Indian class and caste system. Being the daughter of a political activist mother she grew up experiencing various national and international influences-clashes of Western and Eastern ethics, fusion of western and Eastern culture, democracy, communalism, Marxism. A good part of her early life was spent in Kerala with her brother and mother. She too experienced the divorce of her parents in her childhood. She too studied Architecture in Delhi and met her husband there. These experiences by the author herself make the family saga more poignant. Secondly, Kerala which is generally known as an advanced state in India not only in economy but also in education and culture, here is the place where the two social transgressions take place. Thus, Roy's selection of the state Kerala enhances the paradoxical effect of the traditional and orthodox social beliefs, customs and those of the modern as they are dexterously pitted against each other.

This paradox is very interesting and necessary to dig into the psychology of the characters and their homeless status. The family consists of the Estha and Rahel, the twin brother and sister, their mother Ammu, their grandfather Pappachi, grandmother Mammachi, their uncle Chako and their unmarried aunt Baby Kochamma. The family, on the one hand, contains most of the postcolonial features like hybrid identity, cross-cultural interactions on the other hand it believes in the Indian class discrimination and religious division. There are ample evidences of adoptions of western culture, custom and language in the novel. But the family is a part of the Indian patriarchal society that follows the traditional social hierarchy. In Indian society parents' home is not considered a woman's own home. She is supposed to live in her husband's home after marriage. Unmarried and divorcee women are disdained and pitied in the society and considered as burden. The fact that after her divorce Ammu is back in her parents' home strips her of any social position and right. In that sense Baby Kochamma and Ammu both are actually homeless. Seeing their status from postcolonial perspective, when Chako says to Ammu, "what's yours is mine and what's mine is also mine" (Roy 28) or Baby Kochamma with her resentment and fury places Ammu below her in social hierarchy it verifies their cultural hybridization. In Indian society women rank below men and divorced women rank below unmarried women. The interesting fact is that natives learn to take the right to divorce from the British colonizers but do not learn how to tackle with it in the traditional culture. Baby Kochamma prefers the colonizers culture, their language but simultaneously maintains the Indian traditional social hierarchy. Though she considers Rahel and Estha as 'Half-Hindu Hybrids' (Roy 22), she is unaware of the fact that she too is a subject of hybrid culture. Thus the children of a divorcee woman, that is also after an inter-community marriage are considered 'doomed, fatherless wail' (Roy 22) and made to realize that they really have no right to live in their maternal grandmother's house. In that way Estha and Rahel are also homeless. Here the twins are actually struggling with the questions of social and cultural identity. With these psychological conflicts which are actually inherent in the cultural assimilation, they are experiencing the conjuncture of old native tradition and dominant hegemony of the colonial imperialism. When Chako says "They were a family of Anglophiles. Pointed in the wrong direction, trapped outside their own history and unable to retrace their steps-because their footprints had been swept away" (25) or "we're prisoners of War...We belong nowhere" (Roy 26) it actually verifies that experience. Caught in this web; they are unconsciously struggling to find their own social and cultural identity which eventually makes them psychological refugees. Famous French psychiatrist and author Frantz Fanon in his analysis of the psychology of racism and dehumanization in *Black Skin and White Masks* (1952) suggests this dependency and inadequacy of the Black people as they always try to imitate the culture of the colonizer out of inferiority complex. It is the internalization of this inferiority complex that makes the postcolonial subject adore everything western. This is very much evident in the character of Baby Kochamma and the Anglophile Pappachi to whom the definition of modern and civilized is to be western.

Estha and Rahel:

The behaviour and characteristics of the twin characters-Estha and Rahel are quite strange and puzzling if we see from the viewepoint of normal behaviour. Though Roy describes the 'Dizygotic' (Roy 2) twins Estha and Rahel as "a rare breed of Siamese twins, physically separate but with joint identities" (Roy 2), they have totally different personality traits. Later, in their adulthood these differences shape their life differently, but they always cared for each other. While Rahel is childlike and imaginative, Estha is quiet and mature. He "was a quiet bubble floating on a sea of noise" (Roy7). While Rahel always lives in the world of fancy, Estha always sincerely deals with the truth. In the early years of their life they "thought of themselves together as me, and separately, individually as We or Us" (Roy 2). Estha, eighteen minutes older brother of Rahel is a very intense and passionate boy but the shock he gets from the incidents in her childhood ultimately turns him into an unnaturally silent boy. The incident of molestation by the Orangedrink Lemondrink man in the lobby of Abhilash Talkies makes Estha see the world differently. A strange kind of frustration and mistrust about the world was slowly growing into his seven year old mind as he thinks "The Orangedrink Lemondrink Man could walk in any minute" (Roy 93). He begins to believe that, '(a) Anything can Happen to Anyone and (b) It's Best to be Prepared" (Roy 93) Again his journey to the History House confirms to the psychological effects of his previous traumatic experience. This repressed trauma he is suffering from makes him psychologically homeless. The same goes with Rahel whose strange behaviours like decorating heaps of dung with flowers, getting thrown out of school on the charges of smoking and setting someone's hair piece on fire demand psychological analysis. The analysis will definitely contribute to a better understanding of the psychology of Rahel and Estha. They grew up together in the childhood but after the tragic incident of Sophie Mol's death their life began to take different path. Rahel first went to Delhi, from there she went to the United States with her husband Larry McCaslin, but she got divorced and returned to Ayemenem to meet her brother. Anuradha Dingwaney Needham truly calls Rahel "a particular kind of a child who became a particular kind of adult" (Needham 381) who used to remain unaffected by the society's usual norms and ideologies.

Seeing from the post colonial view point the arrival of Sophie Mol, their English cousin slowly makes Rahel more insecure. Baby Kochamma always gives preference to Sophie Mol than Estha and Rahel because of her English origin. Even before Sophie's arrival she started to train Estha and Rahel in preparation of her arrival. "Whenever she caught them speaking in Malayalam. She levied a small fine which was deducted at source…She made them write lines…I will always speak in English, will always speak in English. A hundred times each" (Roy 18). Baby Kochamma is actually the representative of those colonized people who believe that the cultures of Europeans are superior to those of the natives. Fanon in *Black Skin and White Masks* (1952) speaks about this societal fantasy of European racial superiority which has created a sense of estrangement and isolation in the colonized people.

Sophie Mol and Velutha's death and the twin's separation:

The two incidents that make long lasting scar in the twin's mind are the drowning of their English cousin Sophie Mol and the death of the untouchable Velutha. Their involvement in these two acts which ultimately lead to their separation contribute to their psychological anguish. Baby Kochamma constantly tries to generate a feeling of sinfulness in them by calling them murderers. "It's the worst thing that anyone can ever do. Even God doesn't forgive that" again "You know that I know that it wasn't an accident. I know how jealous of her you were…It wasn't an accident, was it?" (Roy 148). They are not even allowed to stand with the rest of the family in Sophie Mol's funeral and everyone seems to ignore them. Baby Kochamma must be partially successful in her manipulative attempt to make them feel guilty. At least Rahel's behaviour verifies that as she never tries to defend herself or deny her aunt's allegation. She gets terrible shock at Sophie Mol's death and Velutha's denouement. She might suspect that her jealous feelings could be the cause of her death. At the same time her imaginative thoughts like seeing Sophie Mol buried alive verifies the fact that Rahel is anguished by her cousin's memories. "When they lowered Sophie Mol's coffin into the ground in the little cemetery behind the church, Rahel knew that she still wasn't dead … Sophie Mol died because she couldn't breathe. Her funeral killed her."(Roy 4) Elizabeth Outka claims in her article that this "Loss' is alive for Rahel at every moment, following her- and even chasing her …from school to school from childhood to womanhood, a frozen moment and yet one that is perpetually on the move."(Outka 27).

Again seeing Velutha beaten to death creates a negative self image in Rahel. Later she also assumes herself guilty of her mother's death as she a thinks it is nothing but the tragic outcome of her maladjusted relationship with her mother. At the time of her death she reminds only Estha considering herself not worthy of her mother's love. On the other hand, Estha to protect his sister, mother and himself becomes a prey to Baby Kochamma's vindictive trick. Sensing her trouble for lying to police she influences Estha to admit that they were kidnapped by Velutha. Right after this incident he is sent to his father whom he hardly knows in Calcutta and after a long gap of twenty three years he is 're-Returned' to the Ayemenem house. In this long period of separation from his mother and sister the connection he shares with Rahel for their each and every dream, thought, feeling and experience is completely cut off. Both of them have travelled aimlessly until they return to their childhood home. Being a victim of the falsehood and cruelty of this society Estha experiences a traumatized childhood. This is where critics bring the theory of trauma. Thus, the two tragic events in the novel, contribute a lot to their psychological state of homelessness. Estha and Rahel might be suffering from what psychological theorists call "Post-Traumatic Stress Disorder" (PTSD) "in which the overwhelming events of the past repeatedly possess, in intrusive images and thoughts the one has lived through them" (Caruth 151).

Social Transgressions:

The two main social transgressions that occur in Roy's novel are Ammu-Velutha relationship and Estha-Rahel incest. By marrying a wrong man to escape from her lifeless life that too by her own choice Ammu has already violated the social rule of arranged marriage. She was living the life of a 'wretched Man-less woman' (Roy 22) not only after the divorce but even before the marriage. Her father, though an Anglophile regarded the college education of a girl as an unnecessary expense. In the Ayemenem house she lived unnoticed waiting only for marriage proposals. "All day she dreamed of escaping from Ayemenem and the clutches of her ill tempered father and bitter, long suffering mother" (Roy 19). Ironically she again fell into the clutches of a "full blown alcoholic"(Roy 19). After the divorce she again returned from her husband's home to her parents' home only to recognize and despise the ugly face of sympathy" (Roy 22). By getting divorced and engaging in an illicit relationship outside the 'Love laws' she makes her situation more dangerous as society expects a divorcee woman to be selfless and sexless. The unfolding of her relationship with the untouchable Velutha ultimately leads to Velutha's tragic death and ejection of Ammu from her family, from the Ayemenem house. Her love affair with the untouchable Velutha against the social norms made her life more difficult and she had to suffer constant humiliation, inside and outside home. After the funeral of Sophie Mol when Ammu with her children went to the police station to give a statement the inspector Thomas Mathew refused to take any statement saying, "Police didn't take statement from veshyas or their illegitimate child" (Roy 5). Ultimately she lost her love and was separated from her kids and died alone in a dirty hotel room. Being banished from her parents' home she was homeless not only psychologically but practically as well. Critics have different opinions about the tragic outcome of this inter-caste love affair of Ammu and Velutha. But there is no doubt that the Indian conservative society here operates as a dominating oppressive force "that saturates virtually all social and cultural space including familial, intimate and affective relationships" (Needham 372).

Coming to the incest of the fraternal twins Estha and Rahel,their breaking down the set social rule or 'love laws' is seen as a healing balm by most of the critics. Their trauma of childhood and anguish and despair of adulthood which make them homeless moving from place to place seem to subside by devoting their body and soul to each other. The psychological refugees Estha and Rahel seem to feel at home when they transgress the love laws by making love. So this transgression can be equated with their progress as far as their psychological recovery is concerned. Giles considers these private (small) struggles of the Ipe family as a mirror of the public (large) identity struggles of the nation. Thus

Arundhati Roy actually portrays the struggle of a decolonized nation to find its own ground mainly through the struggle of the twins and their mother Ammu.

Conclusion:

To conclude, it can be said that the idea of homelessness in *The God of Small of Things* is found in diverse ways. Considering different aspects of this novel both in form and content, like the narrative structure, traditional Indian social structure,transgressions and of course the post colonial issues like hybridity, the term homeless can be used both in its actual and metaphorical sense. In the process of analysis the potential power of psychological trauma to damage an individual's life in relation to the cultural conflict and societal force in a postcolonial nation is established. And as we see the central characters of the novel moving from place to place, city to city and nation to nation sometimes by choice and sometimes by societal force their actual homelessness is also established. The cultural horror and disorder of a postcolonial nation is well brought out through the traumatic memories of the characters. But on a positive note, at the end of the story Roy also leaves a possibility of the recovery of the past traumatic memories by subverting the post colonial social regulation through the union of Estha and Rahel. Possibly they will find their home in the homeland now.

Works Cited:

Caruth,Cathy. "Recapturing the past: Introduction." *Trauma: Explorations in Memory*. Ed. Cathy Caruth. Baltimore and London:The Johns Hopkins University Press,1995.151-157.warwick.Web.22 January 2015.

<http://www.warwick.ac.uk/fac/arts/...13/caruth_recapturing_the_past.pdf>

Fanon,Frantz.*Black Skin,White Masks*.Trans.Charles Lam Markmann.Pluto Press,2008.Web.October 2014.

<abahlali.org/files/_Black_Skin_White_Masks_Pluto_Classics_pdf>

Giles,Michelle. "Postcolonial Gothic and The God of small Things:The Haunting of India's Past." *Postcolonial Text* 6.1(2011):n.pag.Web.2 December 2015.

<http://www.postcolonial.org/index.php/pct/article/download/1192/1108>

Hall,Stuart. "Cultural Identity and Diaspora". *Identity:Community,Culture,Difference*.ed.Jonathan Rutherford. London:Lawerence& Wishart,1990.222-237. Web. 6 January 2015.

<http://www.rlwclarke.net/Theory/...HallCulturalIdentityandDiaspora.pdf>

Longworth,Sarah Y. *Trauma and the Ethical Dilemma in Arundhati Roy's The God of Small Things*. M.Athesis.University of North Carolina Wilmington, 2006.Web. 10 November 2015.

<http://www.libres.uncg.edu/ir/uncw/f/longworths2006-1.pdf>

Needham,Anuradha D. "'The Small Voice of History' in Arundhati Roy's The God of Small Things'.*Inventions:International Journal of Post Colonial Studies* 7.3(2005):369-391.Taylor & Francis Online.Web.18 December 2014.

Rajeev,G. "Arundhati Rai's The God of Small Things –A postcolonial Reading". *IRWLE* 7.2(2011).Web. 7 December 2014.

<http://www.worldlitonline.net/arundhati-rai-s-the.pdf>

Roy,Arundhati. *The God of Small Things*1997.mindpowerindia.org,n.d.Web.2 December 2014

<http://www.mindpowerindia.og/.../MP074_The-God-of-Small-Things-By-Arundhati-R...>

Selden,Raman.Widdowson, Peter.Brooker,Peter,eds.*A Reader's Guide to Contemporary Literary Theory.*5th ed.Pearson Education,2006.Print.

Outka,Elizabeth. "Trauma and Temporal Hybridity in Arundhati Roy's The God of Small Things." *Contemporary Literature* 52.1(2011):21-53.Web.18 January 2015.

<http://www.scholarship.richmond.edu/cgi/viewcontent.cgi?article=1057&context...>

Chapter 3

History or His Story: Searching for an Answer in Salman Rushdie's *Midnight's Children*, a 'Historiographic Metafiction'.

Dipanjan Ghosh
Research Scholar
Department of English
University of Kalyani.
West Bengal 741245.

Salman Rushdie's *Midnight's Children* chronicles Indian history from 1919, the year of Jallianwalabag Massacre, to 1978, the year when the Emergency was lifted. The intervening sixty years of Indian history are interrelated with the rise and fortunes of three generations of the Sinai family. The narration involves the microscopic family nad the macroscopic nation into its account. Rushdie once stated that the novel "can be made to represent many things according to your point of view, they can be seen at the last throw of everything antiquated and retrogressive in our myth –ridden nation whose defeat was entirely desirable in the context of a modernizing Twentieth Century economy". The novel follows the narration and the life of the protagonist Saleem Sinai to weave an alternative narration on the birth of the nation. *Midnight's Children* refers to the birth of Indian nations at the stroke of midnight on the day of Indian Independence.

The narrator-protagonist of Salman Rushdie's novel, Saleem Sinai is the embodiment of a supreme moment of history. As the novel follows the course of his narration and participation through a crystalisation of its evolving mood, retrospection or the distillation of his nostalgic vision, through this fiction, an alternative history is documented. Sinai's narration becomes sometimes critical, sometimes philosophical and somewhat ambivalent. The narration is self-conscious, self-reflexive, even paradoxical and critical of this narrative enterprise. With a project to simplify the vision of reality and make it meaningful, Sinai involves historical documentations, amnesia and historical narration coexist, thereby shattering the coherent grand narratives on the building of the nation. Rushdie accepted the novel as a sort of revisionist reading of Indian history from the microscopic perspective that takes into account undocumented historical and social facts, oral histories, subaltern histories and the contradictions inherent in the dominant history.

Midnight's Children is a panoramic epic on Indian social history passing through the most turbulent periods of nationalist struggle, independence and post-independence struggle for building a nation. As Fielding defines a novel, Rushdie's *Midnight's Children* is likewise an epic in prose that blends history with fiction, fiction with romance, romance with fantasy, and fantasy with history. Unlike historical novels, Rushdie sacrifices historical chronologicity, factual details regarding event, place, action, character, time etc., and then combines fictional narrative with historical narration. The novel presents discontinuous, random, fragmented, even obscure historical events by this experimental mode of narration that rejects chronologicity, logic, or signification. Things are seen as they are, or even as they might have been, or as the individuals think they had been. The plurality of such narrations becomes the theme of the novel. The traditional model of historical discourse with its emphasis on cause, logic, consequence, lesson etc. is replaced by an unauthorized version of Indian history. The historian-as-author, with sufficient authority over the historical text or the narration of nation, is replaced by a subaltern narrator who does not claim any authority over history. The novel uses postmodern narrative strategies to weave a historical narrative from a postcolonial perspective. The colonized 'other' attempts to rewrite the history of the 'self' and 'nation' or 'nationhood' through this unique and experimental novel.

The narration of Indian history in the master-narratives is revisited for revision. Saleem Sinai's narration of a personal history is juxtaposed to the dominant history of the Indian nation. Historical discourse has always favoured the master class. History is an ideology, preserved, propagated, manipulated and stimulated. The authenticity of history is universally acknowledged although plurality and history are not often integrated. In anti-colonial writings, analysis of history has often accommodated multiple histories. Yet, such historical fields are often left unexplored or guarded for the safety of nationalist agenda. History is a method of writing about the 'self', 'nation', 'nationhood'. Like other structures that society founds, history too, is a superstructure shaped on the dominant perception of the base. Post-structuralism introduced a critique of a foundational history for uncovering marginalized histories. Rushdie's novel serves the same purpose. Within fictional mode, Rushdie attempts to rewrite or at least record multiple oral histories revolving around Saleem Sinai. Through the narrator, the subaltern speaks, attempting to link the self with the nation, being born at the stroke of midnight, on August 15, 1947, at the precise moment of India's independence.

According to Ania Loomba:

Histories written from anti-colonist perspectives have re-written the 'story' of capitalist development itself so that the 'grand narrative' of capitalism now appears in a very different light… (Loomba 244)

The postcolonial historiographical narrative attempts to dismantle the structure and create alternative space for subaltern histories. History is dynamic not because the perception of such events or dates change, but because the perception of such events undergoes regular evolution and renewals. History is shapeless and open-ended. In Rushdie's novel, it is marked by fragmentation, selective amnesia about history, even historical solipsism. This recognisation of history as a personal one, often blurred by non-remembrance or the failure of remembering history which makes *Midnight's Children* a metafictional historical narrative. In the novel, we find a continuous engagement with history: a mutual reorganization of the 'personal' and the 'universal' history of Indian nation. According to Peter Hulme, there is a specific reason for moving away from grand narratives because "the grand narrative of decolonization has, for the moment, being adequately told and widely accepted. Hulme proposes that in the present circumstance smaller narratives are much in demand. Saleem Sinai's hi(s)/story is both a fictional 'story' and an autobiography (his story). The backdrop of Indian Independence, or the history of the nation covering a period spanning from 1919 to 1978 is coloured by intense subjectivity and fictionalization. The distinction between history and story is consciously broken by the narrator whose narrative is a continual engagement with the process of making narration and lives.

The epical span of the novel and the multi-layered discourses that incorporate heteroglossia and polyphonic structures allows the novelist to arrive at a configuration of fact and fiction. The documented history is replaced by an individualistic rediscovery of national history. Rushdie tries to strike the key-note of the novel by blurring the distinctions between the public and private histories. The emergence of the concept of nationhood is analysed to shape the new national narrative. In the Indian context, the book appears to be a pastiche of multiple perceptions of Indian reality in the post-independence scenario. The chronological history or biography is replaced by a free-flowing, fluid, both ephemeral and timeless histories of nation. The variegated mosaic of *Midnight's Children* displays a structural and thematic patterning of disparate elements, voices, narratives and perceptions. In the introduction to *Nation and Narration* Homi Bhaba points out that nationalism, by its very definition, is ambivalent. Such ambivalence on the concept of nation is mirrored in every form of national narrative. Saleem's internalization of the Indian national ethos is unique. He personifies India by integrating several of these disparate elements—the ambiguous past, the undefined present and the inexplorable future. The persona and the political converge in this encounter with history. Saleem, the participant, Saleem the narrator and Saleem, the historian, converge at a point of exasperation and inexplicable confusion. His perception of history is modified by his role as a narrator, while his personal self interferes with both. The process of churning of events involves a unity in all its diversity. Saleem is a victim of history, and at the same time, its protagonist, according to Josna E. Rege.

The novel is a personified mixture of multiple histories. C.N. Ramachandran, in the essay "The Empire Lingers on: A Note on the Rushdie Phenomenon" refers to the favourable reception of rushdie's *Midnight's Children* in 1980. He points out that within no time the British and American critics appropriated Rushdie to the "bandwagon of postmodernism and postcolonialism". The critics ignored the aspect of historiographic metafiction, although they accepted Rushdie's experiment with language, narrative, magic realism and metafictional aspects of the novel. History is problematised in the novel and thus ignored by the western critics. Critical tribute from India mainly focuses on Rushdie's engagement with history. Anita Desai in a recent interview (2007) has also acknowledged that Rushdie's innovative treatment of history and narration has opened new scope for experimentation and exploration. There are several references to the acknowledged truth unwritten in the historical text; and there are fictional events concealed by the narrator as historical. Rushdie, according to Steve Connor, can be regarded as a pre-eminent postcolonial writer because his novels "expose the fictionality of history itself". He endorses faith in Linda Hutcheon's view of postmodernist literature as "historigraphic metafiction" and then applies the phrase to Rushdie's novel. Dicter Riemenschneider in "History and the Individual in Anita Desai's *Clear Light of Day* and Salman Rushdie's *Midnight's Children*" (1990) has observed that the novel exposes the fictionality of history "in its absurdity [the absurd story of Saleem Sinai] exposes the extent of the historian's and historiographer's hubris".

The narrator catches the glimpse of past not through the written grand narratives but rather an engagement with the memories of the past as told to him by his ancestors. The narrator talks about his grandfather getting caught in the Jalianwalabag massacre. While describing an eyewitness account of the massacre, the narrative falters:

The fifty-one men enter in the compound and take positions.... As Brigadier Dyre issues a command, the sneeze hits my grandfather fall in the face. 'Yaaaakh—thooo' he sneezes and falls forward, loosing his balance, following his nose and thereby saving his life". (Rushdie 41)

The historical date in this context, the narration of the event, and the oral descriptions are combined in a polyphonic historical discourse. The narrator is also aware of this engagement with different histories. The grand narrative begins in the language of a text book:

On April 13th, many thousands of Indians are crowding through this alley way [way to Jalianwalabag]. "It is a peaceful protest", a person in the crowd tells Doctor Aziz, perhaps an anonymous person who draws Doctor Aziz into the sight of the massacre.

This information, quoted by the narrator, has no authentic historical source. Similarly what the narrator knows about the massacre is partly superadded with fictional improvisations. When the realism and history fails, the narrator takes the refuge of symbolism and allegory. The nose, here, is a major symbol in the novel, and also a comic pastiche. The hierarchy of meaning is collapsed or dismantled by this symbolism. In the opening pages of the book the narrator narrates something thet he has never seen, such as "I was born…on August 15, 1947 when clock hands joined palms respectful greeting as I came…." In the second section, the circular hole, some seven inches in diameter of the bed sheet allows the narrator to see his 'clock-ridden', 'crime-stained' birth. This is the birth of the nation too, seen by the new-born nation itself. India's tryst with destiny has begun, burying the crime-stained past and keeping the appointment with the clock-ridden hour. Such negotiation with personal and the national emerges as an attempt to nullify history through paradox or parody. The narrator refers to history and culture, the popular belief and conception. He uses such expressions that often confuse the readers about the credibility of narration. While describing the highly dangerous form of optimism that his grandfather contracted, he uses ambiguous statement. The inexplicable action of his grandfather, perhaps narrated to the narrator by some old acquaintance is retold. There is no distinction between fact and fiction in such narrations. Even the events of the recent history are not faithfully recorded by the narrator. The narrator fails to recognize important dates and their relevance. There is a conscious effort to undermine history or the new myth about India by exposing the falsity of the dominant myth and history. Rushdie's history can be read as an alternative to the colonial histories of James Mill, Macaulay, Churchill or even Nehru. The re-invention of history is fraught with a sense of shame, fear and anxiety. The hegemony of dominant history persists and fails to shatter the premise on which historical discourse is founded. Rushdie's postcolonial narrative simply replicates a version of history based on personal memory and experience. Important events like the displacement of refugees after the partition, the Bangladesh war of Independence, Emergency or the emergence of Indira Gandhi, the linguistic politics etc. are demystified without any alternative authentic historical documentation. Voice is raised against the mode of historical representation. A dialectics of protest and paradox is introduced in the treatment of history; but like all other postmodern arts, Rushdie's *Midnight's Children* replaces historiography with a pastiche or something hyper-historical.

Works Cited:

Cuddon, J.A. *Dictionary of LITERARY TERMS & LITERARY THEORY.* 13th ed. London: Penguin Books, 1999. Print.

Loomba, Ania. *Postmodernism and Postcolonialism.* Print.

Ramachandran, C.N. "The Empire Lingers on: A Note on the Rushdie Phenomenon." (1992): n. pag. Print.

Rushdie, Salman. *Midnight's Children.* First. London: Vintage, 1995. Print.

Chapter 4

Agony of a Subject: a Postcolonial reading of Chinua Achebe's *THINGS FALL APART* and *NO LONGER AT EASE*

- Debraj Das
Assistant Teacher
Matchpota High School (H.S.)
Matchpota, Nakashipara
Nadia, West Bengal

Chinua Achebe, in his essay *An Image of Africa: Racism in Conrad's Heart of Darkness,* is argumentative to focus on the racism patent in the specific case of *Heart of Darkness* in the following way:

Africa as a metaphysical battlefield devoid of all recognizable humanity, into which the wandering European enters at his peril….The real question is the dehumanization of Africa and Africans which this age-long attitude has fostered and continues to foster in the world. And the question is whether a novel which celebrates this dehumanization, which depersonalizes a portion of the human race, can be called a great work of art. My answer is: No, it cannot.

Achebe, born in the village of Ogidi in eastern Nigeria in 1930, experienced the world of colonialism. Nigeria, as a construction of European colonial powers, was under British control from 1906 to 1960. By the time Achebe was born, literature from postcolonial societies was dwelling upon a new terminology precisely known as 'Black writing' or 'Black literature' and 'Négritude'. To talk about the 'Black writing' model, Bill Ashcroft, Gareth Griffiths and Helen Tiffin, in their book *The Empire Writes Back: Theory and practice in post-colonial literatures* opine, "This proceeds from the idea of race as a major feature of economic and political discrimination and draws together writers in the African diaspora …". The concept of Négritude was the earliest attempt to carve a mould of African writing. Developed by the Martinician Aime Cesaire (1945) and the Senegalse poet Leopold Sedar Senghor, the concept was a firm assertion of the essence of Black culture and diaspora. "Black culture", as the trio say in the former mentioned book, "was emotional rather than rational; it stressed integration and wholeness over analysis and dissection; it operated by distinctive rhythmic and temporal principles, and so forth. Négritude also claimed a distinctive African view of time-space relationships, ethics, metaphysics, and an aesthetics which separated itself from the supposedly 'universal' values of European taste and style."

Achebe seems to believe that Art has a lot of things to do for society. In his famous essay *The Novelist as a Teacher* (1965), which was published almost after eight years of his first novel *Things Fall Apart,* he vindicates, "The writer cannot expect to be excused from the task of re-education and regeneration that must be done. In fact he should march right in front…" *Things Fall Apart,* published in 1957, delineates the African culture along with the superstitions and religious rites through the Ibo society. The novel is a translucent record of the traumatic consequences of the Western Colonialism on the traditional values of the African people. Michel Foucault argues that Power is exercised and employed through a net like organization and individuals not only circulate between its threads but also remain in the position of concomitantly exercising this power. Achebe's *Things Fall Apart* limns how the White colonizers first forcefully introduce an alien form of administration, education and religion and then induces the Igbo people to regard their own religion and culture with derision.

Things Fall Apart depicts Okonkwo, the fictional hero of the novel, as the leader of the struggle against colonial powers. He is the leader of his village. His overt masculinity is a complete contrast to that of his father Unoka. He hated his father who, in Okonnwo's term, was both gentle and idle. To talk about Okonwo's depiction of masculinity, Frank Salamone in the essay titled as *The Depiction of Masculinity in Classic Nigerian Literature* pens, "Masculinity became a metaphor for resistance to those assaults since both culture and society and the indigenous cultures and societies it sought to transform in theory were male-dominated ones. Sexuality, additionally, and eroticism were integral discourses in the colonial discourse, symbolizing power and control." Needless to say that this masculinity would be a hindrance to the white colonizers who would come to Umuofia.

In a heterogeneous society, postcolonial writers usually try to reassign new ethnic and cultural meanings to the groups of people who are treated as insignificant by the society. As a postcolonial novel, *Things fall Apart* describes both the perfections and shortcomings of Igbo culture to differentiate it from the Western culture. So, throughout the novel we see that the narrative is very comfortable at the religious belief and superstitions of Igbo people. They had an array of Gods and deities ranging from the personal God "Chi" to the Supreme God "Chukwu". Other deities like Udo, Ogwugwu and Idemili were believed to protect their clan and culture. Their asinine behavioral pattern is well sketched by Achebe in their superstitions. Thus we find Igbo parents advising their children not to wrestle at night for fear of evil spirits and "A snake was never called by its name at night, because it would hear" (*TFA* 8). The twitching of an eye-lid is considered as a bad omen: "It means you are going to cry" (*TFA*, pp.30). If a person had swelling in his stomach, it was considered "an abomination to the earth goddess. When a man was afflicted with swelling in the stomach and the limbs he was not allowed to die in the house" (*TFA*, pp.14). The people also observed a "week of peace" before sowing seeds since, it was believed, goddess Ani would get pleased and this would lead the village to prosperity. The belief in the omens and ill-omens was so deep in their psyche that it had become instrumental in shaping their behaviour and conduct: "'Is that me?' Ekwefi called back. That was the way people answered calls from outside. They never answered yes for fear it might be an evil spirit calling" (TFA, pp.30). The birth of twins was considered a bad omen and they were put in earthenware pots and were mercilessly thrown away in the forests. "The Oracle of the Hills and Caves" of Umuofia eventually "pronounced" that Ikemefuna, who considered Okonkwo his second father, must be killed; and one fine morning the execution was done in which even Okonkwo took part: "...Okonkwo looked away. He heard the blow. The pot fell and broke in the sand. He heard Ikemefuna cry, 'My father, thay have killed me!' as he ran towards him. Dazed with fear, Okonkwo drew his matchet and cut him down. He was afraid of being thought weak" (*TFA*, pp.44)

The question of 'Otherness' comes at the second part of the novel when "The Missionaries had come to Umuofia. They had built their church there, won a handful of converts and were already sending evangelists to the surrounding towns and villages" (TFA, pp.105). The emergence of new religion did become successful in creating some doubts in the minds of the village folks. The young generation felt attracted towards the new religion called Christianity. "The arrival of the missionaries had caused a considerable stir in the village of Mbanta" and the white man took it as his burden to teach the villagers that "…they worshipped false gods, gods of wood and stone" (TFA, pp.106). The white missionary also felt it poignant to teach the black people that the god worshipped by them (the black people) were not gods at all as they were the gods of chicanery who would tell them to kill their fellows and destroy their innocent children for the sake of religion.

Soon, the new religion captured the some of the Igbo people. It was Okonkwo's first son, Nwoye, who had been captivated by the hymn if the new religion. He could never accept the killing of Ikemefuna and the hymn of Christianity, especially "[t]he words of the hymn were like drops of frozen rain melting on the dry plate of the panting earth." He was, as a result, greatly enthralled. The missionaries needed a small piece of land to begin with. The rulers of Mbanta, another village like Umuofia, gave the missionaries an 'evil forest' since the rulers believed "in it were buried were all those who died of the really evil diseases… An 'evil forest' was, therefore, alive with sinister forces…" (TFA, pp.109). Ironically, this marginal forest in which the missionaries would set up their church would, later on, marginalize the Igbo people. "One of the most striking contradictions about colonialism is", as suggested by Ania Loomba in her book *Colonialism/Postcolonialism*, "that it needs both to 'civilise' its 'others' and to fix them into perpetual 'otherness' (pp. 145). The new religion welcomed everybody since it had the burden on its shoulder. The Igbo people, especially the women, who were somewhat terrified and tortured so far by their male counterparts sought refuge to the shelter of the missionaries. So, Nneka had given four childbirths but could not enjoy the feeling of being a mother because each time she had borne twins, they had been immediately thrown away since twins were inauspicious to the Igbo people: "Her husband and his family were already becoming highly critical of such a woman and were not unduly perturbed when they found she had fled to join the Christians. It was [rather] a good riddance" (TFA, pp.111).

Homi Bhabha believed that 'Resistance' is a condition which is produced by the dominant discourse itself. Okonkwo decided to resist the emerging missionaries. He felt deeply hurt to see his village and his people changing and breaking away from their values and beliefs. His Nationalism coloured his anti-colonial rage: "He had shaken out his smoked raffia skirt and examined his tall feather head-gear and his shield. They were all satisfactory, he had thought" (TFA, pp.145). He wanted Umuofia to go to war against the missionaries but he knew nobody would protest against them. In a fit of anger he killed one of the messengers of the District Commissioner Court, who came to stop the meeting of the villagers in which Okonwo was present thinking that the meeting could be the last attempt to resist the colonizers. But after the execution was done, Okonkwo found that the villagers "...had broken into tumult instead of action. He discerned fright in that tumult."

Okonkwo, the brutal wrestler, had to wipe his matchet on the sand and went away. He committed suicide since he perceived that the colonizers had crept into the blood of the Umuofia people. One may doubt the success of this novel since the hero commits suicide. But Achebe had a different treatment. Having shown Okonkwo's dangling dead body to the District Commissioner, Obierika said to him, "Perhaps your men can help us bring him down and bury him" (*TFA* pp.151). The Commissioner, appalled as he had become, asked them why they themselves could not do it. Then came an answer which, I think, marks *Things Fall Apart* as a postcolonial novel: "It is against our **custom**... It is an abomination for a man to take his own life. It is an offence against the Earth...His body is **evil**, and only **strangers** may touch it. That is why we ask your people to bring him down, **because you are strangers**" (*TFA* 151) [emphasis mine]. Achebe showed that their customs and superstitions would still continue and this continuation would be the mark of the existence of Igbo culture although Okonkwo would be "buried like a dog."

Now let us talk about Achebe's *No Longer at Ease*, the second novel of our consideration. Published in 1960, *No Longer at Ease* basically depicts how education, as an ideological state apparatus, was used to colonize African people. According to Althusser's theory, ideological instruments like education make the colonized nations accept the power of the ruling class. The novel was published in the year of Nigeria's independence from England. It was like a shifting pattern in Nigeria when the Whites were leaving and the Blacks were getting responsible for their own discourse of life. It was time for them to rediscover their diaspora. Colonialism was giving way to a postcolonial situation and the Nigerians were trying, amidst such a ruffled time, to find out a way of negotiation between the claims of colonial modernity and the previously degraded mode of life.

The title of the novel seems to be inspired from T.S.Eliot's poem *The Journey of the Magi*. The wise men brought gift to the baby Jesus Christ. After enduring hardships, these Magi returned home and they seemed unimpressed by the infant. The birth of Jesus marked the end of Paganism. The Magi returned to their countries but they were no longer at ease there. Achebe probably chose this title because Obi Okonkwo, the protagonist of the novel, not only got alienated from his own country (Nigeria) but also could not get himself acclimatized to the culture of England, the country he studied in. He is entrapped in the dialectic of difference and identity. Leela Gandhi, in her *Postcolonial Theory: A Critical Introduction* talks about the theory of Homi Bhabha who, as she says, "...announces that memory is the necessary and sometimes hazardous bridge between colonialism and the question of cultural identity... It is a painful re-membering, a putting together of the dismembered past to make sense of the trauma of the present" (*Postcolonial Theory* pp.9). Having completed his studies in England, when Obi arrived in his country, he went to Lagos; the city which was placed between Europe and Umuofia and was closer to Obi's heart. His first impressions of Nigeria were formed in England but after his arrival he was disillusioned: "It was in England that Nigeria first became more than just a name to him... But the Nigeria he returned to was in many ways different from the picture he had carried in his mind..."(*NLE*, pp.11). He had no idea about the slum of Lagos. The city was turned into an African-European city. There was no darkness because "at night the electric shines like the sun". The city had lost its African tradition and was divided into two parts. One part belonged to the upper class people; the majority of whom held European posts like Obi. And the counterpart was the slums. Although Obi was brought up in a culture in which solidarity and tribal life was of prime importance, his education in England inculcated the value of individualism and self-will in his mind. Education as a colonialist instrument turns English into a norm, while internalizing in the mind of the colonized his inferiority. Obi's father, inspite of being an old African man, believed in the spell of the words. He himself had a good library. He knew that the power of the White men was in their printed book and the words written in them. The Africans had their own material too for writing but that faded over time.

The people of Umuofia wanted Obi to study law. Although he studied English instead of law, his European degree and his post in the Civil Service (the highest rank below a European) could come to their help too. Ironically, he learned nothing about the law and ended up as a victim of law in the court: "Treacherous tears came into Obi's eyes. He brought out a white handkerchief and rubbed his face" (*NLE,* pp.2). The colour of the handkerchief is significant. The idea is that the Umuofians wanted him to study law in order to pursue the language and literature "colonizer". But the people of Umuofia did not know that Obi's education would bring alienation. His ideas and beliefs were changed and he no longer was at ease to accept his traditional life. Memory, Homi Bhabha believes, is the submerged and constitutive bedrock of conscious existence. While some memories are accessible to consciousness, others, which are blocked and banned, oscillate the unconscious in several ways: "The procedure of analysis-theory, recommended here, is guided by Lacan's ironic reversal of the Cartesian *cogito,* whereby the rationalistic truth of 'I think therefore I am' is rephrased in the proposition: 'I think where I am not, therefore I am where I do not think'" (*Postcolonial Theory,* pp.9). Similarly, "Obi found that his mission house upbringing and European education had made him a stranger in his country" (*NLE,* pp.64-65). Obi's clan did a great deal to send Obi to Europe. Obi knew that. What the kinsmen did not know was "that, having laboured in sweat and tears to enroll their kinsmen among the shining élite, **they had to keep him there.** Having made him a member of an exclusive club whose members greet one another with 'How's the car behaving?' did they expect him to turn round and answer; 'I'm sorry, but my car is off the road. You see I couldn't pay my insurance premium.'(*NLE,* pp.90)? [Emphasis mine]. Obi had been given the burden of bringing light to his dark village. His trip to England represented a new form of fulfilling traditional expectations.

Bribery is one aspect which has been discussed a lot in the novel. The first question Obi faced while going to opt for the Civil service job was, "Why do you want a job in the civil service? So that you can take bribes?" The interesting thing was that taking bribes actually caused fewer problems than refusing it. Obi was placed in a dilemma. On the one hand he could not make his ends meet anymore and on the other his determination for an honest life did not let him accept bribes. But he had to succumb to the situational irony. Finally when he was offered the bribe and the money was kept by the anonymous man on Obi's table, Obi "...would have preferred not to look in its direction, but he seemed to have no choice. He just sat looking at it, paralysed by his thoughts" (*NLE,* pp.153).

Obi finally succumbs. He succumbs to the colonial power. But Achebe seems to project the idea how the colonized is defined by the ruling power and is derived not from his own rights but from respect also. Achebe was aware of the tools that were used in colonizing African people. But Achebe makes the novel stand triumphant by locating the loopholes in Western education system. After returning to Africa Obi had tried to be useful and help his people, but he could not do that. It is not surprising for a person like Mr. Green to say, "The African is corrupt through and through"(*NLE,* pp.3); but we are to stand beside Obi since he debunks the myth that "a man in need of a job could not afford to be angry [to his colonial masters]" (*NLE,* pp.37). And the novel vindicates the rights of the African people when Achebe employs Obi to say, "It is not the fault of Nigerians. You devised these soft conditions for yourselves when every European was automatically in the senior service and every African in the junior service. Now that a few of us have been admitted into the senior service, you turn round and blame us." (*NLE,* pp.140)

Bibliography

1. Achebe, Chinua. *Things Fall Apart*: Penguin Modern Classics.
2. Achebe, Chinua. *No Longer at Ease*: African Writers Series: Heinemann Educational Publishers.
3. Gandhi, Leela. *Postcolonial Theory*: Oxford University Press.
4. Loomba, Ania. *Colonialism/Postcolonialism*: the New Critical Idiom Series.
5. Ashcroft, Bill; Griffiths, Gareth; Tiffin, Helen. *The Empire Writes Back*: Routledge, London and New York
6. The beginning excerpt has been borrowed from *The Norton Anthology of Theory and Criticism*: Norton.

Chapter 5

When Color Imperialism Prevails over Literary (*The Bluest Eye*) and Digital Media: From Post colonial View-Point

Sriparna Chakraboty
M.A. (English), B.ED.
Guest Lecturer
Chakdaha College
Chakdaha, Nadia, W.B. India

*A*bstract: - *From the very dawn of civilization, different colors have been playing different significant roles in human lives. Specially, the colors- black (dark) and white are normally taken together as fixed in a binary oppositional mode. So far no problem is arising until and unless one color (white) dominates the other one (black/dark) and badly exploits its existence. This condition attracts immediately our eyes as these two colors are inextricably tied up with us both physically and mentally. These colors bear different connotations attached to our very mundane humane lives. Color black or dark is "always already" there to symbolize anything impure, bad, evil or negative; whereas the color white, as if, takes the burden to purge all those things associated with the former one. Thus it symbolizes purity, good, or something very positive. We know very well that once our loving earth was born out of the dark and chaotic dancing of universal masses. Then it was lit with gaseous rays. Everything that happened set against dark backdrop as that very dark setting gave birth to this earth with all its glitz. So how can this particular color be tagged with anything cheap and devilish? And who derogate this color on such an insulting base? These questions can be delved best if we look into the different artistic manifestations down the ages more or less chronologically. Now a days, the very form of media has a mirroring effect on common mass. Media as powerful instrument stir our lives with inevitable truths from our surroundings. If electronic media is taken as digital one, literature, newspaper, journal, hoarding with poster* *etc. can be treated as printed media. All these are the open delineation inviting us to discover those bedded answers and much more.*

Keywords: - *Post colonialism, post modernism, racism, subaltern, queer, imperialism, apartheid.*

"There can't be anyone, I am sure, who doesn't know what it feels like to be disliked, even rejected, momentarily or for sustained periods of time. Perhaps the feeling is merely indifference, mild annoyance, but it may also be hurt. It may even be that some of us know what it is like to be an actually hated — hated for thing we have no control over and cannot change."(Foreword, *The Bluest Eye*)

It is not always we talk about 'black beauty' and how that kind of beauty soothes our eyes from the scorching rays of white. Rather, now a days, we are far more engrossed ourselves into the packets of beauty kits for getting a fairer new us. Those who have already been boarding into this ever fairing and brightening ship, are feeling blessed with their accomplishment. And the unfortunate ones, in spite of real hard works, who cannot be able to have even a minute change or a lighter tone on their skin, start cursing their beauty products as the products fail to curve a desirable imprint on their customers' skin as well as on their minds. So now the question arrives, who has set or who will set what is desirable for everyone? Is it a homogeneous world that same thing will go with everyone's interest? And finally, who will judge or justify whether a particular person can attain his/her desirable thing or not? This is the high time and we should start anchoring or thought into these questions in order to take out the real sense of being situated on double standards of our society.

From the dawn of civilization, brought in the hands of colonizers, in that period of colonization, the colonized people were treated "non-European as exotic or immoral other". Specially the colonized ones with brown and black skin were taken as inferiors as creatures, who badly needed the proper training and teaching in order to get shaped into that mould of their superiors' shadow. As 'white man's burden' can be unburdened by themselves only so for perpetrating this so called 'noble mission' of civilizing brute people from developing 'third world countries', European white masters started preaching "Children, both black and white,…to see history, culture and progress as beginning with the arrival of the Europeans." Thus they sowed the poisonous seeds of colonialism at the very beginning in the 'tabula rasa' of children's minds, so that they can never be able to question the 'white empire' in order to assert their own native identity and to search for their pristine roots. Post colonial writers and critics have already taken out the colonial reasons with which, "for centuries, the European colonizing power will have devalued the nation's past, seeing its pre-colonial era as a pre-civilized limbo..." The colonized mass started looking for recognition from authority at the cost of their pure ethnicity. They were compelled to emulate the values set by the superiors. Thus, from that socio-political compulsion of colonial era, a latent desire of becoming fairer just like Europeans comes to writhe the minds of black and brown people. The more they get fairer, the more they can attract both similarities and authentication from 'West'. A long phase of oppression by Europeans in colonial period has created a strong impact over the colonized minds. Though they have arrived into a post colonial world of individual independence, they cannot shirk the values of colonizers. They still follow those dictums of Westerners psychologically. The horror of racial marginalization, specially based on color-politics, looms large in the psycho-geographical existence of each and every person who has got comparatively brown or black skin. On this very ground different devices or product are manufactured to recuperate them out of this fear of getting aberrant. So these persons are out to try their fairness cream to get satisfied mentally. They do not want to be hated as 'other', coming from a 'surrogate' world. Sometime, these terrified people want much more to have for adorning themselves as per with the latest fashion of this world. They just keep imitating the poise of those they presume their superiors and idols and so earnestly hankering after the possessions of those superiors. Thus they, somehow unknowingly fall into the trap of power-construct of this hegemonic society and they are crushed under the wheel of domination and exploitation.

Such is the case with the protagonist of the novel *The Bluest Eye (1970)* by Toni Morrison. Pecola Breedlove, a foster child, who finds a shelter into the home of 9-year-old Claudia MacTeer and her 10-year-old sister Frieda in Lorain, Ohio. Claudia and her sister live with their parents and after Pecola's house getting burned down by her wildly unstable father, Cholly, their parents somehow become kind enough to add Pecola to the family. It is a complex world of love-hatred-faith-doubt-desire-passion-craze-submission, webbed the individual minds with awe. Claudia, at the very outset makes it clear that unlike her sister Frieda and Pecola she is not going to adore any white existence. Being a poor girl, torn between society's dubious standards based on the ground of racial discrimination, she submissively voices her dislike for white-blonde beautiful dolls, ladies and little girls. She just grudgingly asserts to herself-

"I couldn't join them in their adoration because I hated Shirley. Not because she was cute, but because she danced with Bojangles, who was *my* friend, *my* uncle, *my* daddy, and who ought to have been soft-shoeing it and chuckling with me. Instead he was enjoying, sharing, giving a lovely dance thing with one of those little white girls whose socks never slid down under their heels….What I felt at that time was unsullied hatred. But before that I had felt a stranger, more frightening thing than hatred for all the Shirley Temples of the world…."

Through Claudia's deep expression Morrison tries to project the racial tension especially with color issue between Black and White in America. We can understand the pain that twitches young minds as they are not being loved by society as they are inborn Black. This stuffy situation gives birth too many questions and doubts, as Claudia herself tries to be getting through these ordeals-

"I destroyed white baby dolls. But the dismembering of dolls was not the true horror. The truly horrifying thing was the transference of the same impulses to little white girls. The indifference with which I could have axed them was shaken only by my desire to do so. To discover what eluded me: the secret of the magic they weaved on others. What made people look at them and say, "Awwwww," but not for me?"

It needs immediate attention how much detestation springs in Claudia's mind out of this painful situation that she is not treated in the same way a white girl being treated by society. But characters like hers have to accept the truth ungrudgingly once they are entering the phase of womanhood. Pecola, the meek, poor, black girl with far more tempestuous life can not reject the societal hierarchical strata and standards personally; rather she is in love with those white beauties. She surreptitiously longs for having such a lovable beauty in herself. So is the condition of Frieda on her ground of individual desire, as narrated by Morrison-

"She was a long time with the milk, and gazed fondly at the silhouette of Shirley Temple's dimpled face. Frieda and she had a loving conversation about how cu-ute Shirley Temple was."

Pecola gradually learns to be a woman to have a body to be loved to give birth to child. There is a fleshy realization about love as so called inferior black woman normally is not granted to have perception of soul through rational senses. Society hybridized with white as its manipulative power-class does not want to understand anything related to the rights of these ethnic marginalized souls. As Bell Hooks in her *'Postmodern Blackness', in Yearning: Race, Gender, and Cultural Politics* (1991, pp.23-31) asserts that the interest and urge of a black person cannot be empathized by the white people. It is a post modern world of free game, of social inversion and of topsy-turviness, and everybody shares a common ground, though the very sense of blackness, as observed by Morrison-"… like a black on the street, already guilty, already a perp" cannot be effaced from the mental arena of white individuals. This the glaring irony that retrieves an innocent and meek mind like that of Pecola who tries earnestly to grasp the all glamorous face and a pair of blue eyes in her body. But the sheer fact of regret and shame is that, this quest for transfiguration brings a heavy blow to the young, passive girl in the nasty evil form of her own patriarchal community. She meets her predicament in the very entity of her own lout father who gives her a sway from this mundane existence of agony to a world of deluded imagination, by raping her. Finally reaching that wonderful psycho-geographical territory, the poor girl, Pecola finds a blissful state in which it seems to her that the pair of blue eyes has ultimately conjoined her mortal body. In order to get the hitherto unattainable blue eyes Pecola has to leave this world and thus can escape its hierarchical grinder and has also to undergo the aching phase of giving birth to an illegitimate baby. She, thus, symbolizes all those scapegoats around us, who clandestinely tread the earth with a sense of guilt, and who are used to make the society look more beautiful and fairer as on the ground of petty comparison. So to stay away from this heart-wrenching journey, we keep us aware of all those cosmetic products which whisper to us "make-up" as it does not seem to be "made-up", by noticing the lengthy advertisements on huge hoardings or on television. Morrison was also concerned with these pangs of the black women-"I was deeply concerned about the feelings of being ugly."

The subalterns of other developing countries, except all those black torn souls, are facing the same tension. Here the situation is becoming worse day by day, as the power-wheel smashes these lives on the ground of casteism and pecuniary possession. European, especially British people looked down upon India as a country of wild animals, snakes, snakes-charmers, savage natives, witch-craft or black-magic etc. and a barren land of intellect. This consensus of foreign people once wrecked the mind with a negative force elite class of this country. Though, once, there had been the rich *Varnasrama* social divisions, based on qualities and work as if- some people have the qualities of a *Brahmin* and if they work as a *Brahmin* they are accepted as a qualified *Brahmans*. This system, soon, fell prey under a corrupt "caste" system in India. This malpractice of caste system gave birth to an oppressed group of untouchables. Besides these, British masters, the opportunists, using the narrow fissures in a multi-cultural and multi-racial country like India, where different ethnicity, religions etc. live together and are strongly interconnected with one another, apply their various provoking issues and conditions which ultimately devastate any possibility of peaceful terms ensuring the harmonious living and unity in the diversity of India. Thus, threats towards the economically, racially, culturally marginalized people are increasing age by age. Women are soft ground for fast targets. The oppression on untouchables and dalits in India always reminds one of racial oppression and segregation in America, South Africa and many other parts across the world. If we go through Bama Faustina Soosairaj's autobiographical novel *Karukku* (1992), we can witness how atrocities pin dalit women as well as dalit men. Imperialism, in its every form like- religion, caste, culture, economy, race, color etc. aggrandize its power over all types of socially backward people. These people also have their own longing to get accepted by the social orders, which prioritizes, in present days, the looks and money. The fair look is one of the must-needed criteria and sometimes it is needed with proper power exertion- money. But these unfortunate people with their dusky skin and tattered lives fail to meet those criteria and thus are left as unaccepted or unrecognized. The negative forces of color imperialism bow our minds such a low stature that influences our opinions harmfully when we are about to choose life-partners for marriage. Here, comes the necessity for a good fairness cream in order to lighten our skin tone for passing smoothly the marriage ordeal. Such is the order of hypocrite world. It is responsible for crushing the mental tenacity and self-confidence of those poor creatures roaming on this cruel earth. It also endangers their lives in this society. One might frequently hear black people are being burnt to death after others accusing them of engaging in witchcraft. In Tanzania or other African country are prone to this type of inhumane activities. India also shows this kind of loathsome violent acts towards socially segregated poor women.

Day by day, society thus is stylized differently with latest weapons to nurture its imperial offshoots in the deep breath of human beings. Across the world, many intellectuals voice against all these negative and dehumanizing instrumentations. Many writers like the South African Nobel Prize winner Nadine Gordimer, with her historically and politically charged novels tries to emphasize the humanitarian cause for braving mean social marginalizing process of 'apartheid', a tyrannical system of racism in South Africa. Cinema, a far more moving and living genre in the field of art also depicts the setting of a new voyage, in this new era, beyond the clutches of imperialism and slavery, by appropriating the voice of black humans. One of such glaring examples of this is the film called, *Twelve Years a Slave* (2013), a slave memoir, discovering a new found freedom of a life bruised by slavery.

Coming back to the mainspring of this discussion- though the world of Toni Morrison's novel *The Bluest Eye* is a paradoxical and ironical one, it explores a new reality hitherto unknown to so called 'normals' among us. The climactic trajectory shows the inner revolution of a poor black girl, wistfully longs for a beautifully loved life decked with a pair of blue eyes, blonde hair and fair skin. She can compete with the hassles of this mundane life and also completes her journey. And after being brutally raped by her own father, she finally finds a confinement for her own freedom, in a mesmerizing world of fantasy. Now she atones for us- as "We were so beautiful when we stood astride her ugliness. Her simplicity decorated us, her guilt sanctified us, her pain made us glow with health, her awkwardness made us think we had a sense of humor. Her inarticulateness made us believe we were eloquent. Her poverty kept us generous. Even her waking dreams we used—to silence our own nightmares. And she let us and thereby deserved our contempt. We honed our egos on her, padded our characters with her frailty, and yawned in the fantasy of our strength." She simply steps up into a madness which turns out to be a blessed state for her also- as it "protected her from us" and our world full of pangs. She is now permitted to have all which once snatched away from her by the step world. Though this feeling sanctifies her but the same pricks us perpetually.

We become wise enough to realize- "it was the fault of the earth, the land, of our town. I even think now that the land of the entire country was hostile to marigolds that year. This soil is bad for certain kinds of flowers. Certain seeds it will not nurture, certain fruit it will not bear, and when the land kills of its own volition, we acquiesce and say the victim had no right to live. We are wrong, of course, but it doesn't matter. It's too late. At least on the edge of my town, among the garbage and the sunflowers of my town, it's much, much, much too late."

Work Cited:

1. Morrison, Toni. *The Bluest Eye,* vintage international, Vintage eBooks, A Division of Random House, Inc. New York.
2. Rice, Philip and Waugh Patricia. *Modern Literary Theory A Reader,* Arnold, A Member of the Hodder Headline Group, London and Co-published by Oxford University Press Inc. New York.
3. http://ehlt.flinders.edu.au/projects/counterpoints/PDF/A14.pdf.Accessed on 10 February, 2015. Web.
4. Duvall, John N. *The Identifying Fictions of Toni Morrison: Modernist Authenticity and Postmodern Blackness* (2000), Palgrave Macmillan.
5. http://www.steppenwolf.org/_pdf/studyguides/bluest_eye_studyguide.pdf. Accessed on 15 February, 2015. Web.
6. http://draper.fas.nyu.edu/docs/IO/4398/Postcolonial.Theory.pdf. Accessed on 5 February, 2015. Web.
7. http://en.wikipedia.org/wiki/12 Years a Slave (film). Accessed on 5 February, 2015. Web.
8. http://www.distinguishedwomen.com/biographies/morrison.html. Accessed on 2 February, 2015. Web.
9. Barry, Peter. *Beginning Theory: an introduction to literary and cultural theory*, Second Edition, © Peter Barry 1995, 2002 ISBN: 0719062683.
10. Ashcroft, Bill and Griffiths, Gareth and Tiffin, Helen. *The Post-colonial Studies Reader,* Routledge, Taylor & Francis e-Library, 2003. London and New York.

Chapter 6

The Voice of Protest: Lallitambika Antherjanam's *Praticaradevatha* ("The Goddess of Revenge").

Kajal Sutradhar

Asst. Professor

Nahata Jogendranath Mondal Smriti Mahavidyalaya

Nahata, North 24 Parganas

West Bengal

Literature represents the multilayered and inherent cultural crisis of society from time immemorial. Lallitambika Antherjanam (1909-1987), the well-known Malayalam writer points outthe oppression of women in Namboodiri society and the blatant double standard of sexual morality in her novels and short-stories. In Women Writing in India (vol-1), Susie Tharu and K. Lalita point out the speech of Lallitambika as she had addressed in a seminar on feminism and literature, "I am very glad we can have a discussion like this today. It should have really happened a hundred years ago.……..It is good that we can, at long last, talk about so many more of which we still cannot speak".

In fact, most of Antherjanam's best stories deal with the helpless condition of women confined in the mudupadam or traditional household. She herself belonged to the Namboodiri: community,- the powerful feudal aristocrats in Kerala famed for their stern adherence to tradition. In her award-winning novel, Agnishakshi (witness by fire, 1976) the writers tells the story of two women who rebel in different ways against the soul-destroying restrictions of the life laid down by their society. In "Confession of Guilt", one of Antherjanam's most powerful stories, the protagonist is the victim of an exchange marriage. Antherjanam's well-known collections of short stories are AdyatheKathakal (First stories), 1934, TakarnaTalamura (Ruined Generation) 1949, Kalivadilued (Pigeon Hole) 1955, and Agni Pushpanjal (Flower of Fire) 1960. In this article,

I will concentrate on Antherjanam's well-known short story Praticaradevatha (The Goddess of revenge),in which,too, the recurringtheme is the callousness of society to the desire of a woman and her desperate attempt to take revenge against the society which speaks of morality but actually is utterly hollow at the centre.

As Susie Tharu and K. Lalita said in woman writers in India, vol: 1, "Lalithambika was born into a family of writers deeply involved in the early 20thc movements for the reform of Namboodiri society. The family home 'which was the treasure house of books and Journals, functioned as a forum for healed discussion in which many important writers of the time participated. Lallithambika was the first girl from the Namboodiri community who dared to wear askirt and blouse." She took education at home and began to write at the age of fourteen. The freedom fighters in colonial India, specially Mahatma Gandhi inspired her a lot. She wrote a short piece when Gandhiji was in jail and sent it to publishers. Thus writing career of Antherjanam began.

Marriage with NarayanaNamboodiri in 1926 transplanted Lallithambikafrom a free intellectual into a closed custom-bound household of Namboodiri family. The placidity and uneventfulness of life could not curb the indomitable spiritof Antherjanam. She herself said "when the door of the outside closed, the door inside opened.........I saw many things at close quarters. I listened, touched, and felt crying without tears, life without breath, rooms in which no blood was spattered but within which, not human beings, but shadows and statues moved. Their smiles and tears were alike."

The way Antherjanam handlesthe role of housewife and a successful writer is very praise worthy. While her farmer husband lived with the hoe and she lived with the pen. She had seven children. She made a harmoniousbalance between her literally life and family life and she never gave in to the acute family problem of her life.

In this article, I will specially concentrate on Antherjanam'sPraticaradevatha (The Goddess of revenge) in which the central figure Tatri: who comes in midnight dreams of the narrator. The story was writer in 1938 and she based on a historical figure KuriyedathuTatri who was symbol of fallen woman in Namboodiri society. It was even sin to speak about Tatri who was a prostitute and at the trial she disclosed at sixty four names who used her privatelyat the same time denouncedher publicly. The sensational incident was reported in "The MalayalaManorama" and "The Deepika" on 5 june,1905.

Tatri appears in the midnight dreams of the narrator with "an intense fire of revenge burned fearfully in her eyes". As the narrator explains, Tatri isawoman........not a young girl......not old either......sorrow, a certain austerity,disgust, disappointment: all these mingledin her expression. "Tatri pours her heart to the narrator. She recollects her days of youth when she fasted on all auspicious days. She made garlands of Karuka and uttered "Parvati Swayamvaram". When she was seventeen or eighteen, she got married to an unknown person, as the narrator sags in the story: "Whether our hands are placed in those of an old man or a young one, a sick man or alibertine, is all a matter of destiny. We can do nothing but endure". This endurance-this helplessness is the destiny of Indian womanfrom time immemorial. They are forced to get married. Their material happiness entirely depends on fate.

Like any other girl, Tatri started her married life with a 'boundless sense of happiness' as the sags: "I did my almost to satisfy his preferences in our conjugal life".........after all, a husband is considered to be a God in person."

In spite of all attempts to please her husband, Tatri become an ignored wife by her husband. Her husband advised her to learn a harlot's tricks to satisfy her husband. It is the age-old Idea that prevails in Indian society that a woman should be like Draupadi in cooking, like a sister in managing the household affairs and like a prostitute to become anidealbed partners to her husband. Tatri adjusted with this Idea and she mastered the artof a prostitute. Her tremendous despaircame when her husband married for the second time: "It happened without any warning: one evening,he came home with his new wife........that was the first time I thought of man as devils." When Tatri cursed the new wife as harlot, her husband used the same words: "I brought her home deliberately, knowing she's a harlot. I like harlots. Why don't you become one yourself?"

A husband telling his chaste, highborn wife to be prostitute petrified Tatri. Immediately her despair turned into anger as she says: "An irrational, uncontrollable desire for revenge took hold of my mind." She returned to her father's housewhich was a journey from the "frying pan into the fire" in the language of Tatri. An innocent desire for the admiration of the opposite sex still lurkedin her mind: "There were men who met my eyes, returned my smile. After all, people tend to smile if you smile at them. It soon became a habit."

Scandalous reports began to spread. Tatri's mother cursed her as she had been born to ruin the family's honor. Her brother's wife did not allow her to enter into the kitchen any more. Tatri was captivated in a helpless situation as she had to take a decision to become a prostitute in order to handle the awkward situation as she said: "I had made my decision. If this was my ultimate destiny, I must transform it into an act of revenge. I must avenge my mothers,sisters,countless women who had been- weak and helpless…."

From then onward, Tatri described herself a passionate and fascinating woman whose bewitching loveliness attracted men to her. She princes, titled chiefs, noble man of all ranks crowded around her. In this context, Tatri says: "The fame of this new harlot spread far wide. These who came to her went away gladdened".

In one night Tatri's husband came to his wife. It was Tatri's moment of reckoning. His words "go and learn to be a prostitute" still alive in Tatri's mind. Before their parting, her husband told Tatri that he had never seen such a passionate and intelligent woman in his life. Tatri replied that was the lie. She told him to remember his wife as the narratorsays: "Light was dawning on him. He looked suddenly at my face, screamed and got up. "oh God! Is it Tatri! Tatri! Tatri!".

This incident stirred Kerala to it's very foundations. From great prince to highborn Brahmin, men trembled because they knew that this harlot might betray their names. Tatri got many gifts from these men such as ring, or golden waist chain. Sheproduced all these to prove the guiltof sixty-five men, including scholars well versed in the Vedas. At last Tatri demands equality of judgement from a common humanitarian point of view as she asks the narrator: "Tell me, sister! Tell me! Who is more culpable: The man who seduces a woman in order to satisfy his lust for flesh, or the woman who transgresses the dictates of society in an attempt to oppose him?".

Here the narrator falls in adilemma. She is sympathetic to the pain of Tatri. At the same time, she did not support her way of revenge. The narrator's voice seems a little bitorthodox when she renounces Tatri'sindividual attempt of protest: "Why did you shoulder the burden of revenge all alone, In matters of thiskind, sister, individuals cannot triumph…..consider,now,what good did it do to society, that hurricane you set in motion?…..Noteven the women in the families of the sixty- fivewho were excommunicated have been released from their agony."

The writer is timid to support Tatri because Tatri's case was unique in colonial Kerala. The writer renounces Tatri she took the revenge alone. Rather she should do it collectively. From the evidence of history, we knew that always protest comes individually. When there is no collective attempts of protest, a single voice is enough to create a storm. Here lies the success of Tatri: Although the whole Namboodiri society considered her 'sinner',-a fallen woman, hers is a single voice protest against the patriarchic Indian society. So, Lallimbika Antherjanam'sTatriin "Goddess of Revenge", though written in colonial era, is still relevant in present time for her courage, intelligence, spirit of protest and altitude to society.

Acknowledgement

i.   Women writing in India. Vol.1',
     600 BC to the early 20th century (Oxford University Press).
ii. Translation of the original story by Gita Krishnankutty (OUP).
iii. Feminist Psyche in World Women Novelists by N.ShanthaNaik (Sarup
     Book Publishers Pvt. Ltd.)

Chapter 7

MAQBOOL: AN ADAPTATION WITH AN ATTITUDE

Monikinkini Basu
Doctoral Research Scholar
Centre for Studies in Social Sciences, Calctta
Jadavpore University

If the thought around third cinema as a concept is made clear then we would find it to be a world order that the focus has been on the re-inventing and recovering of local aesthetics and traditions of narrative as against an international patterning of the set Hollywood order. It also explores the ways in which the suppressed internal others of the nation, whether of class, sub or counter nationality, ethnic group or gender can find a voice. In the representation of a Shakespearean play what becomes vital is the fact that the essence of "Indian-ness" remains intact in the form of the typical Bollywood style of looking at it.

"Aj tera pandrah August hai"... "kya sahib...November mein pandrah August kaha se ayega" ("it is 15th August for you today"... "What are you saying,how will it be 15th August in November)..." tujhe Mughal ke bistar se chutkara jo mil gaya" (since you are independent of being Mughal's keep now). The very first scene of the film centres on the fact that a homosexual man who was the head of an underworld family gets killed and his keep gets killed as well. The person who survives the deal is Boti, the son of Mughal. Set in the backdrop of cultural loci of the Indian film industry, *Maqbool* complicates the issue of popular culture from its very inception. The fact remains that what is portrayed in the film is considered as the norm is the first place. The culture that is represented in the film is that of oppression and self doubt. Both of these factors contribute to the formation of a genre in films that deal with enormous self doubt. The other important aspect of the film is the focus on the two policemen.

The two men replace the witches in the play, and therefore necessarily fulfil certain criterion that was exclusive to the witches. The fact that they remain outside the circle of the underworld and are far off from the context of law as well, make them a misfit in all areas. They are old, marginal and probably just an escape route for the goons and never can occupy the centre. Their approach to the family structure is marred by the fact that they are not at the centre but are very prominently marginal. The presentation of the witches in the form of two men belonging to the police department is very much an example of subversion of stereotypes since the police is supposed to be the agents of Repressive State Apparatus; here they function as the agents imbibing the minds of the characters with the concept of Ideological State Apparatus.

In Vishal Bharadwaj's representation of *Macbeth*, it is interesting to note that the method of adaptation is unique to a local setting. The Mumbai Underworld very elegantly replaces the Feudal set up of Scotland and then some characters are modified to suit the necessities of the edited plot. The very first change that is noticed in the film is the replacement of Malcolm and Donalbain by a single character Sameera who is called Choti by Abbaji, the Duncan figure in the film. This replacement becomes significant since not only one character is presented as a replacement of two, but the gender has been altered as well. The replacement continues as Nimmi replaces Lady Macbeth and is designated as the keep of Abbaji(the Duncan figure), instead of being the wife of Macbeth, (Maqbool in the film). But the most important replacement is that of the witches by the police inspectors, Purohit and Pandit (played by Naseeruddin Shah and Om Puri). The treatment of their characters is perhaps the most interesting aspect of the adaptation.

The role of the witches in the play has been pivotal to the plot, but differing from that in the film the role has been made very complicated in the sense that the police men have assumed the role of the director of actions in the film. It is through them that the narrative is anticipated and the characters act according to their game plan. The repetitive usage of the sentence, "Sakti ka santulan, bohut zaruri hai, aag ke liye pani ka darr bane rahna chahiye" which can be loosely translated into "It is of vital importance that the power equation is maintained, for fire, the threat of water is necessary", points out at the activity of the policemen which was like an equilibrium that maintained the underworld ferocity and helped to keep gang-wars at bay. The idea of the two policemen replacing the witches can be seen as a transgression of the popular idea of witches, which is that of old women who have been through their menopausal period and thus the hormonal imbalances have resulted in their having a body which is useless to the society (since, female body is the site of procreation, once that ability is lost the body becomes unwanted). Their roles, however, in the film have made them transform from an ordinary puppet in the hands of Abbaji to the actual mastermind working behind the power game that topples the throne and replaces the "King of Kings" in the blink of an eye. But if we look at the scenario in another way, we will find that the two policemen can be seen as homosexual individuals who are probably on a mission to take up arms against the Heteronormative world around them. If this consideration is made of their role, then we have to admire the intensity of their patience with which the control and coordinate the fate of the Mumbai underworld.

Their first appearance in the film and the drawing of the chart depicting the fate of Mumbai is significant enough that the future of the state lays in their prediction, or rather their predicaments. This dissolves in, to make the face of Maqbool appear on the frame, therefore validating that it would be he who will be the tool to control the Mumbai-Underworld. It is they who pass the information of the whereabouts of Mughal and Boti (gang men from another underworld family) and thus Maqbool becomes the reason for the death of Mughal and therefore gives Boti (the Macduff figure, who has been spared the treachery by Guddu) a reason to hate him. The first and the most important provocation for Maqbool was planted when at the farmhouse Pandit claims that Maqbool would be getting the leadership of "Bollywood", and obvious lucrative prospect for someone from the underworld, since glamour and wealth are in abundance in the Mumbai Film Industry. Pandit goes on to claim that Maqbool shall be replacing Abbaji within six months. The reaction of Maqbool and Kaka (the Banquo of the film) are noteworthy, since Maqbool threatens Pandit with death and Purohit supports him claiming that Pandit has an evil tongue providing example of successful prophesies that have been shocking to him. Kaka however changes the topic and asks Pandit to assess his fortunes instead. Pandit configures on him that he should have been dead by then and that his son, Guddu (the Fleance in the film), shall be the check to Maqbool's rise to power. This scene can be ideally called a parallel of the first scene of the first act of the play. But what is interesting is the location of the scene. Instead of the barren heath that is tormented by the rough weather and the witches, it is Maqbool's home that serves as a location for this instigation and therefore the locale becomes equally tainted and ominous as the barren heath. The cast off landmass suddenly is relocated just into the very existence of Maqbool's dwelling place. The proximity of Pandit and Purohit to the household vouchsafes the threat that they possess for Maqbool and the upcoming usurping of the throne of Abbaji becomes even more pronounced. The duo is almost gate crashing at every family union and therefore the statement "aag ke liye pani ka darr bane rehna chahiye" becomes almost synonymous to their presence which is almost forever in the family. This prolonged influence is serialised in the fact that the predictions are done by one of the two and the other makes sure that the prediction makes a mark on the person for whom it is made. When Maqbool threatens Pandit against his predictions of Maqbool being the one to replace Abbaji, Purohit makes sure Maqbool believes in him and offers examples to support the prophesy. Even Abbaji's decision to make Maqbool the in-charge of the Hindi Film Industry is cited as an example to Maqbool, of what lies next in line is even greater success. The complexity of the two characters increase with the advancement of the plot and therefore the fact that the centre of power shifts from the gang into the hands of administration is very clearly portrayed in the film. Pandit and Purohit, may seem to be the so-called puppets in the hands of Abbaji, but the game plan that they have is symbolic to the one that popular culture always has for its exponents. The replacements are so fine and invisible to the naked eye that one never realises the concept of mirage that is so intricate to anything popular.

The prophesy-scene in *Macbeth* is replaced in *Maqbool* only to further contribute to the theatricality and emotional angle of the film. The greater question that arises here is of individual choice and fate being a provocation used to politicise the context of the action. The policemen play a very choric part in predicting for the audience as to what will follow in the narrative. They have the perfect sense of humour in mentioning facts at the opportune moment. When the duo arrive for the "mangni" (engagement) of Choti without any gifts, they conclude that what they have brought is "subh samachar" (good news) for them, and which is that Devsare, the ACP who tried to bring the underworld under control by arresting Abbaji has been transferred to Customs

Department and shall be on sea forever. This however does not bring much happy memories for Maqbool since he was publicly insulted by Devsare. In the mind of the already dwindling Maqbool, the relocating of the fact that unlike Laljibhai (the Godfather to Abbaji), who had killed a policeman who had insulted Abbaji, on the day of his marriage, Abbaji hadn't "punished" Devsare, thus making him reconsider the fact if at all serving Abbaji is worth it. By ensuring the prospects of their trustworthiness on Maqbool, the "witches" of the film actually turn up side down the original project of Shakespeare. The 'outer' chaotic world of the witches in *Macbeth* was the 'other' for the staunch feudal setup of the 'inner' arena of Scotland. But in Maqbool, the equations are completely reversed as the chaotic underworld becomes the 'other' which is intersected by the exponents of law, without the sound of their footsteps.

The hugely wronged character in the film would certainly be Nimmi, who falls prey to her love and is very well utilised as instigation by the policemen when the need arises. They take no time in conferring the guilt of Abbaji's murder on Nimmi with lines "ab kamre mein teen log the..Abbaji, Usmaan, aur..." (there were three people in the room, Abbaji, Ussman and...) and they end the line with a laugh, as if the third person Nimmi had intentions which were obviously known to be vulnerable for Abbaji. Obviously, the mistress of an underworld figure has always been seen as a plotter and conspiring persona, and Nimmi is politically made the same by the policemen. They even let Boti escape and join Guddu to form forces against Maqbool. The question of their intensions being the cleaning up of the gangs of underworld is answered by them in one of the initial scenes when Devsare questions them about their whereabouts and they claim to be on duty, and feel proud of the fact that without much effort one of the gangs got perished and the other is under their control. Even though the tone here was that of false claim, they actually could very well be credited of doing the same.

The prospect of using the two policemen as the agents of the Repressive State Apparatus has been obvious to the director and hence they act as the eye of surveillance for the otherwise worriless household. Their mode of executing their social responsibility is unique in its own way. They remain the "faithful friends" of Maqbool, thus confirming his tragedy and the associated downfall of Abbaji's empire. Pandit and Purohit, the actual plot makers who feel empowered since they are so much well camouflaged in the colours of the underworld. It would never occur to a person without knowledge of their actual intentions that these two men have been gaining power with every scene of the film. In the last section of the film when Pandit forbids Purohit to devour 'Shani' (Saturn) in his fortune teller chart, then Purohit asks Pandit as to what exactly would happen if he ate the Saturn of the prediction, to which Pandit says, 'Shani insaan ko khata hai, insaan shani ko nehi' (it is Saturn that devours human beings not the opposite) and adds, 'aur ajkal yeh bohut bhukha hai' (it is very hungry these days). Purohit asks Pandit, 'kise khaega?' (who is to be devoured) to which Pandit answers, 'kise khilana hai?' (name whom you want it to eat). Thus the power that they enjoy in the course of action of the film is something that they realise themselves to be capable of handling. This essentially is very revolutionary. Since the witches in the play *Macbeth* have never been subjected to such extensive self evaluations and they have never made their own roles so pronounced, even though the play confirms them of holding preternatural powers. In Maqbool, the conditions change very dramatically. From being the men in action they gradually give themselves a face lift to become the king makers. The power of the underworld is famous in the social set up. But what really works behind the power is the politics of popular culture. The fact remains curious as to how Bharadwaj makes the plot exceedingly intriguing when he uses the characters without much screen time. The tool that Bharadwaj uses is obviously the most vulnerable side of popular belief, astrology. The name Pandit in Hindi means the learned one, and Purohit means one who performs the rituals. Both these names are associated with a Hindu temple or Hindu religious prodigies. It is very interesting to note that in the film Pandit offers the 'prophesy' while Purohit convinces everyone of its truth. This combined effort creates havoc on the minds of the other characters and therefore it becomes easy to trample with their minds and confer on them the consequences of their actions.

The idea of murdering Abbaji is also served by Pandit when he narrates the tale of Abbaji coming to power by killing Laljibhai, the godfather of Abbaji. But there remains a lot of doubt about the truth of the statements uttered by Pandit, even though Maqbool is made to believe the words, the audience is not much convinced of it. Therefore, the two men amount to what Stuart Hall addresses as "organic intellectuals" (Hall 1788), that is people who know a little more than other intellectuals and are more profound in their knowledge. The amount of faith that the two policemen have on each other is something that needs to be focussed on. They never get disharmonised or disoriented. They function primarily as an institution and not as an individual. This qualifies them as the marginalised centre of the plot and steadily they craft the 'fortune' of the underworld.

What also works very steadily in the film is the politics of race, culture and religion. The fact that Guddu is finally held as the next ruler of the underworld can be seen as a politicised activity of the two kingmakers, who make it a point to confer the laurels of their achievement on someone from their own religion. It may be seen as a complimentary aspect of popularised notions of the eternal quest between the Hindu and the Muslim for a position of superior ability to control situations. Here as well, it is no different. But however with Nimmi's child being born, the quest remains unfulfilled.

Thus, Bharadwaj politicises the situation, making the plot critical in the sense that the politics that popular culture plays on the minds of the audience and the characters has to undergo various levels of dissociation. The framing of the witches in the garb of policemen, the outer being deeply inner to the plot, makes the film a complex kaleidoscope of Indian cinema, covered under the mask of entertainment, but leaving the audience with a quest in mind.

But if we consider the other representation of the policemen as mentioned loosely in the beginning of the argument, that is as the queer individuals who are perhaps in a relationship with each other (since for most of the film they are seen in company of each other) and the fact that being gay has been a reason for their marginalisation then probably a different idea would spring up to end the argument. They can thus be read as the socially deprived who try and bring up depravity for the socially succulent. The target for them will be the heterosexual men who are in positions of power. These men are made the tools of the policemen. They destroy the procreative society and bring down the powerful to their mercy. Thus they convert the cycle a complete 360 degree and make the power equations swing. Their entire planning can be seen as a meditation against the social hierarchy that perhaps has made them the outcasts. Thus the popular culture that binds the narrative can be seen as a nuance which makes up the personal political and vice versa. This reading probably makes up a partial understanding and leaves much to the audience to dig into.

Works Cited

Hall, Stuart. "Cultural Studies and Its Theoretical Legacies". Rep. *The Norton Anthology of Theory and Criticism*. Ed. Vincent B. Leitch. New York: Norton. Second Edition, 2001. (1782-1795).

Chapter 8

THE UNTRODDEN EXPANSE OF THE SUBALTERN PSYCHE IN DORIS LESSING'S *THE GRASS IS SINGING.*

By Soma Das
MA (Eng) B.Ed
Ex-part-time Lecturer in English
(Rashtraguru Surendranath College, Barrackpore)
Present Designation: Assistant Teacher
Raghavpur Jr. High School
Panpara, Nadia

Introduction

The Grass Is Singing (1950) is Doris Lessing's first novel, which according to Ruth Whittaker is "an extraordinary first novel in its assured treatment of its unusual subject... Doris Lessing's questions the entire values of Rhodesian white colonial society (28). Lessing shares some of her experiences and memories in this novel on her upbringing, childhood, and youth as a white settler in the Rhodesian veld. The novel portrays eloquently the author's disapprobation of the political prejudices and white colonialism in the Southern African setting through the life of Mary Turner, a white landowner's wife and her ruinous relationship with her native servant. This article discusses in full length the white settlers' behavioural pattern moulded by the conceptions of class, gender, and race and executed through post-colonialism, and the reaction of the subaltern placed in the midst of imperialism.

The black/white binary

"Mary Turner, wife of Richard Turner, a farmer at Ngesi, was found murdered on the front verandah of their homestead yesterday morning. The houseboy, who has been arrested, has confessed to the crime. No motive has been discovered. It is thought he was in search of valuables".

The novel starts with this piece of a news not uncommon in the Rhodesian society of the 1950s and the reaction of the white settlers are "a little spurt of anger mingled with what was almost satisfaction, as if some belief had been confirmed". The people who are bearing the colour black on their skin are truly the symbol of darkness and evil and this bit of news registers the fact. "When native still, murder or rape, that is the feeling white people have". However the white colonizer's survival is incomplete and unaccomplished without the black natives as they are their forced farm labourers. Lessing eloquently says "they, (the natives) the geese that laid the golden eggs, were still in that state where they did not know there were other ways of living besides producing gold for others". However, as the novel proceeds, all such hatreds and ruthless exploitations on the natives culminates in the portrayal of the protagonist, Mary Turner. Life was different for her before her hasty marriage to Dick Turner. "She had never came into contact with natives before, as an employer on her own account. Her mother's servants she had been forbidden to talk to…'native problem' meant for her other women's complaints of their servants at the tea parties. She was afraid of them of course". This fear for the unknown native with colour, speech, and cultural differences builds up an abjection among the whites. Unlike the other white mistresses, Mary is unable to suppress her hatred and this results in the consecutive dismissal of the native servants. After Samson (the old native working for Dick for a long time) gives up the job due to the Mary's misbehaviour "came a native to the backdoor, asking for work. He wanted 17 shillings a month. She (Mary) beat him down by two, feeling pleased with her victory over him". However, this boy further increased her irritation and hatred as "it made her angry that he would never meet her eyes. She did not know it was part of the native code of politeness not to look a superior in the face, she thought it was merely further evidence of their shifty and dishonest nature". Lessing seems to be very clear in her protest against colonialism while she handles these lines, that the white settlers, not only undermines and looks down upon the natives while imposing on them the burden of their culture they most inhumanly bother not to understand and pay consideration to the native culture as well. Mary seems to epitomize the colonial lords. "If she disliked native men, she loathed the women. She hated the exposed fleshiness of them, their soft brown bodies and soft bashful faces that were also insolent and inquisitive. Above all, she hated the way they suckled their babies with their breasts hanging down for everyone to see". After their six years' of marriage Dick got ill for the first time and Mary decides to go down to look after the farm, "she had to crush down violent repugnance to the idea of facing the farm natives herself". She with an air of an employer carries the long forgotten sjambok with more confidence of and vindictively she clings to her idea of the blacks as "filthy savages!". Her supervising the farm however, makes the natives even more repulsive and hateful, however, the natives worked like machines with their half naked thick muscled black bodies stooping in the mindless rhythm of their work. The abjection surpasses all limits while a native exhausted with work and heat asks for water and that too in English which was unbearable for the white lady (Mary) as speaking the employer's tongue was an offence and to Mary a deliberate conspiracy to insult. Maddened with anger "involuntarily she lifted her whip and brought it down across his face in a vicious swinging blow". According to Jan Mohamed, "the native is cast as no more than a recipient of the negative elements of the self that the European projects onto him". The enslaved black natives whether as a domestic servant or as farm labours thus can be represented as wild, violent, potential rapists and always threatening to the white women.

Race/Gender/Abjection

As Julia Kristeva says abjection is a desire to expel but powerlessness to achieve it. It is "directed against a threat that seems to emanate from an exorbitant outside or inside, ejected beyond the scope of the possible, the tolerable, the thinkable". (Kristeva 2). Through abjection, Mary establishes herself as pure and good and considers her racial "other" as impure and disgusting. Mary attempts to negate her sexuality and femininity which are rooted in her childhood trauma but she does not have the self consciousness to analyze and recognize them. She watches the native women who have humane drives of maternity as "others" as the site of "abjection". The reason that leads Mary to execute her rage and hatred against the native is her denial and repression of sexuality. Much of her unconsciousness are projected on the natives, both the domestic servant and the farm labours. She thus considers them as "black animals" with "a hot, sour animal smell".

The Final Intervolvement

Mary's already complicated life with a frustrated marriage, poverty, isolation, and fear of sexuality intensifies with Moses' entry in her life as a houseboy. Moses is the same worker whom Mary struck with a whip two years back. She is "unable to treat this boy as she had treated all the others for always, at the back of her mind, was that moment of fear she had known just after she had hit him and thought he would attack her. She felt uneasy in his presence"(p142). Yet there is an element of sexual attraction in her towards the houseboy, the powerful broad-built body fascinates her. Once Mary sees Moses half naked washing himself; he stops and stands upright with his body "expressing resentment of her presence there" (p142) waiting for her to go. Mary is filled with historical anger "that perhaps he believed she was there on purpose this thought of course was not conscious, it would be too much presumption, such unspeakable cheek for him to imagine such a thing..." (p143). The formal pattern of black and white, mistress and servant has been broken "by the personal relation" (p143). She now sees in Moses, a man with "the powerful back stooping". However, Mary is aware of the colonial rule that: "When a white man in Africa by accident looks into the eyes of a native and sees the human being (which it is his chief preoccupation to avoid). His sense of guilt, which he denies, fumes up in resentment and he brings down the whip". (p144). Mary is threatened by the man in Moses who affects her both sexually and culturally. She feels she must do something to restore her pose and immediately asks Dick to dismiss the boy. Dick however, tired with the endless dismissing of servants insists that Moses should stay. Mary slowly looses her balance with the fact of being alone with Moses in the house and becomes, in Ellen Brooks' words, "pray to violent emotions, which she can neither understand nor control, stemming from deeply embedded psychological repression". She feels "one of a strong and irrational fear, a deep uneasiness and even though this she did not know, would have died rather than acknowledge - of some dark attraction". When Moses wants to leave the job instead of unleashing her violent rage as was common to her, she sobs in front of him and Moses controls the situation "like a father commanding" (p152). She feels "helpless in his power" (p154) and resigns her authority. She is sexually attracted towards him which is unthinkable, a taboo. She is both afraid and fascinated by him, "a terrible dark fear" (p152). Michael Thorpe notes Moses intrudes "not as mere symbol of colour conflicts but as the agent of a disruptive life force" and triggers Mary's long-repressed emotions to act out her traditional female role, helpless and dependent on him" (12). She is attracted towards his strength, energy and grace which is a sharp contrast to Dick's "lean hands, coffee-burned by the sun", which seems to be trembling and weak. Even Moses' act of kindness and caring and his desire to please surpasses the unwritten law concerning the relationship between the black and white. Mary is racially dominant by psychologically and sexually dominated Moses. However, Lessing emphasizes that "the white civilization will never, never admit that a white person, and most particularly, a white woman can have human relationship, whether for good or for evil, with a black man" (p26). Michael Thorpe remarks that: "Since 1903, in Rhodesia, it has been a criminal offence for a black man and a white woman to have sexual intercourse but no such law applies where a white man and a black woman are involved" (12).

The subaltern does speak

"People did ask, cursorily, why the murderer had given himself up. There was not much chance of escape, but he did have a sporting chance. He could have run to the hills and hidden for a while. Or he could have slipped over the border to Portuguese territory..." (p12-13).

The houseboy does not run away after committing the murder of his mistress, neither does he plea for his life as Lessing says "well, it was the tradition to face punishment and really there was something rather fine about it...although, it is not done to say things natives do are 'fine' ". But for Moses it was a silent protest against the colonizers who lovingly and dolefully and satisfactorily draws a conclusion that he was actually in search for the material 'valuables'. No moral values can belong to them. They are the subaltern mass suffering a position without identity existing beyond the hegemonic power structure of the colony. Moses' first encounter Mary Turner was in the agricultural field where she embodies the true spirit of a colonizer well equipped with a sjambok armed with resentment and hatred which has no bounds unlike Dick who well handles the 'native problems' with his suppressed amount of hatred for them. What Lessing is really showing is how damaging the colonial psyche can be when one is not equipped for it and Mary is surely ill equipped for it. Equal amount of resentment is cast for the whites as Dick says 'they say we stink' (p115). This resentment gathered high when she hysterically whipped Moses for speaking English and asking for water. "A thick weal pushed up along the dark skin of the cheek as she looked and from it a drop of bright blood gathered and trickle down and off his chin, and splashed on his chest" (p119). However, Moses knew that Mary was frightened "she saw him make a sudden movement and recoiled, terrified" his expression "turned her stomach liquid with fear" (p120). Two years after when Moses was appointed as the houseboy "she felt uneasy in his presence". However, he made no suggestions that he remembered the incident. "He was silent, dogged and patient under her stream of explanations and orders. His eyes he always kept lowered, as if afraid to look at her". However, things changed and Moses assumes greater and greater importance in Mary's life while she "faded, tousled, her lip narrowed in anger, her eyes hot, her face puffed and blotched with red" can hardly "recognize herself" (p146). Moses could understand that "Mary moved about her work like a woman in a dream". The displacement of anxiety about the degeneration of her position as a colonizer and lifelong repression places Mary into madness. Moses witnesses this not like a machine but as a human being and takes care of her 'almost fatherly'. However, Mary's constant ill treatment of him disheartens him and he says "I do the work well, yes...then why Madam always cross?" However, with a passage of the time a new familiar friendliness builds up and Moses questions about the work and much to the surprise of Mary, he asks an universal potential question almost bordering on the merits of social thinkers, "Did Jesus think it right that people should kill each other?". The astonishment of Mary was however catered by Dick that Moses was a 'Mission boy' and that was still another reason for the whites to dislike the mission boys because they 'knew too much'. In Dick's second fit of illness, Moses could feel the helplessness and frustrations of Mary and took the situations at hand to look after Dick while Mary can take some rest. He could even read Mary's mind that beyond a dominant mistress there is a clinging fear that she bears of him and thus like an assuring friendliness "he said easily, familiarly why is Madam afraid of me?". These uneasiness creeps away as time moves on heavily and slowly and Mary's suppressed fear of sexuality makes her physically as well as mentally dependent on him, so much so that she breaks code of the colonial rule, the 'esprit de corps' and has a discourse with him in the most unfamiliar way even before Charlie Statter much to his shock, and it was quite evident as Dick says 'Mary likes him' (p178). This incident gives a new twist. Violating the colonial code is a disease that needs a thorough treatment. Charlie Slatter, the representative of the true colonizer forces Dick to give up his farm and work as a paid manager and makes him understand the seriousness of moving away and leaving the place immediately. "You should get off at once. You must see that for yourself". Moses was a silent witness looming from his hidden hide. It was unbearable for him now that he was mentally and physically attracted to Mary. Situation worsens with the appearance of Tony who "was twenty" and "had a good, conventional education". Tony sees clearly the attraction and repulsion between Moses and Mary, particularly at a crucial time when he witnesses just by chance Moses helping Mary to dress and buttoning her with an attitude of "an indulgent uxoriousness" (p185). When Mary sees Tony there, she is scared and by the intervention of another white man, Mary reacts negatively towards Moses and dismisses him, while Tony is putting his arms around her shoulders to comfort her. Tony realizes that Mary is asserting herself and using his presence "as a shield to a fight to get back a command she had lost". The shock of betrayal was quite naturally unthinkable on the part of the colonizers that Moses suffer.

"Madam want me to go?" said the boy quietly.
"Yes, go away".
"Madam want me to go because of this boss?" (p188).

Moses goes away and to Mary's knowledge will return back revenge. She knew what she did, it was a blow that did cut much deeper into the mind of Moses unlike the blow of the whip which left just a scar. He again returns back to us.

"Madam is leaving this farm, yes?"
"Yes" said Mary faintly.
"Madam never coming back?"

...

"And this boss going too?"

Moses gets his answer and vanishes in the dark and then looms in the darkness awaiting for the right moment to avenge his betrayal. Since the subaltern has not a voice to speak out his heart, his emotional conflicts, his demands and mostly his betrayal, he can only take revenge and bring upon the poetic justice. His protest lies in his confession and surrender.

"Moses himself rise out of a tangled ant-heap in front of them. He walked up to them and said, (or words to this effect): 'here I am' ".

He is no more in a position without identity - as a subaltern should have been. However, the irony lies in the fact that the colonizer has the power to distort or even falsify the truth and thereby justifies themselves and portrays the dehumanized dark picture of the black natives as evil.

Conclusion: A blurred tragic hero

Moses does grasp his own identity by taking his due revenge, and thereafter surrendering himself. He is not like a common black. He is a mission boy with better education and character. He is the only black whom Mary has to respect. He is not afraid of her, rather most sympathetically he understands her frustrated calamity and tries to please and takes care of her. This leads to a sort of affection and possessiveness for her. However, on contrary, Mary does not show one iota of empathy for him, her fascination for Moses was because he is something exotic. It was born out of utter desperate solitude, not of any genuine affection. Moses was tricked by this. It was not with the interference of tony that he should feel the betrayal, rather he was betrayed from the very beginning of the interfacing affair with Mary. However, this is nothing but the deep seated racism which proves the colonizer not to consider the natives as fully human. However, though apparently, the storyline seems to be a bildungsroman of Mary, the story is actually of Moses larking behind and feeling each gap. It begins and ends with him, his lost identity is finally evident as he commits the murder, fulfils his revenge and slips into the speechless silence of protest. "Though what thoughts of regret or pity or perhaps wounded human affection were compounded with the satisfaction of his completed revenge, it is impossible to say. For, when he had gone perhaps a couple of hundred yards through the soaking bush, he stopped, turned aside, and leaned against a tree on ant-heap. And there he would remain until his pursuers, in their turn, came to find him".

Works Cited:

Brooks, Ellen W. *Fragmentation and Integration: A Study of Doris Lessing's Fiction.* Ph.D. thesis. New York University, 1971.

Kristeva, Julia. *Powers of Horror: An Essay on Abjection,* trans. L.S. Roudiez. New York: Columbia University Press, 1982.

Lessing, Doris: *The Grass is Singing.* London: Flamingo, 1994.

Mohamed, Abdul Jan. "The Economy of Manichean Allegory: The Function of Racial Differences in Colonial Literature" in *Race, Writing and Differences.* Ed. Henry Luis Gates. Chicago: University of Chicago, 1985.

Maslen, Elizabeth. *Doris Lessing. Writers and their Work.* General Editor: Isobel Armstrong.

Thorpe, Michael. *Doris Lessing's Africa.* Harlow: Longman Group, 1973.

Chapter 9

Amitav Ghosh's *The Shadow Lines:* Indian Writing in English from a Post-Colonial Perspective

Jayini Ghosh,
Guest-Lecturer,
Dept. of English, Kanchrapara College,
North 24 Parganas, W.B.

The Indian Writing in English[6] has a long history of its own which is closely related to the history of British colonial rule in India. English was introduced in India with the advent of the British East India Company in the sub-continent around 1612 and was made the medium for formal education and important government works around 1835 when Macaulay's minute was given the approval by Lord William Bentinck. This marks the beginning of official bilingualism in India. Macaulay wanted to create a bourgeois middle-class, Indian by race and colour but British by taste and principles, apt to work as interpreters between the colonial masters and the native people. This would help the handful of the British rulers to rule over the vast Indian territory. Macaulay realized that English will not only serve an immediate purpose, but will have a far reaching effect as well. Long after the British had left India, the language of the masters will remain with the people as an indelible mark of their colonial subjugation. The word colonialism originates from Roman 'colonia' which meant 'farm' or 'settlement'. The *Oxford English Dictionary* defines it as —

A settlement in a new country... forming a community subject to or connected with their parent state; the community so formed, consisting of the

---

6   From now on Indian Writing in English will be referred to as IWE in this essay.

original settlers and their descendants and successors, as long as the connection with the parent state is kept up.

Ania Loomba rightly points out that this entire definition does not refer in any way to the colonized people, the violence and oppression endured by them and the forceful encroachment of land and power in a foreign land. This definition makes it appear as a smooth process which involves only the colonizers.[7]

The introduction of English in India however opened up the avenues of western philosophy and science before the native people. Under the leadership of intellects such as Raja Ram Mohan Roy, Ishwar Chandra, Rabindra Nath Tagore, Swami Bibekananda, Mahatma Gandhi, Aurobinda Ghosh, Bal Gangadhar Tillack and many others the youth of India were liberated from their orthodox outlooks and prejudices. Thus inadvertently Macaulay and Lord Bentinck actually provided the Indians with a means of liberation not only from the clutches of foreign rule but also from the ill practices of superstition, caste discrimination and unfounded prejudices.

The very first generation of Indian poets and authors in English includes the likes of Michael Madhusudan Dutt, Tagore, Aurobindo Ghosh, Mulk Raj Anand, Raja Rao, R.K. Narayan, etc. They were tremendously inspired by the Victorian and Georgian writers. This was a generation closely following the western norms and principles in the field of culture and creativity, constantly infusing Indian tradition with the western derivative discourse. Over the ages IWE went through various stages of development. However it was around 1980s that the IWE achieved a truly universal approach. We find bold experimentations in the thematic aspects as well as in the narrative techniques. Indian authors in English have become masters of magic realism, reinterpretation of history from a personal point of view and multi-laired narration, etc. These authors have challenged the stereotypical western ideologies which they feel have become either stagnant or inadequate to represent life in literature any more.

The new generation of authors includes stalwarts such as Salman Rushdie, Amitav Ghosh, Arundhoti Roy, Anita Desai, Kiran Desai, Jhumpa Lahiri and many more belonging to the Indian diaspora. In this paper we will closely

---

[7]   The discussion about the meaning of the term colonialism and Ania Loomba's views about it are taken from the book titled *Colonialism/Postcolonialism*, chapter one, page 1, by Loomba.

focus on Amitav Ghosh's *The Shadow Lines* and try to analyse it from the Post-colonial perspective. The Post-colonial era began with the termination of the European, predominantly the British colonial rule over the rest of the world. There is no fixed date which may be referred to as the beginning of the Post-colonial age. However with the end of World War-II, there was a definite decline in the colonial powers of the west. There was an alteration of the male domain of activities, from the adventurous outer world to that of nation building, something quite similar to house mending, a task usually reserved for the women.

In *The Shadow Lines* Ghosh has presented certain important literary and social concepts in a new light. The most important among these concepts is perhaps the idea of "self" and "other". From around the sixteenth century, when the Europeans had first started their explorations all round the globe, the western world considered the white, masculine, Christian, men to be the representative of the "self". Even the white women were considered to be the second race or the second grade citizens — the "other" Amongst the obvious differences, they were believed to be lacking in logical rationality. The orient was a region of mystery and magic devoid of all logic and rationality. In places like Africa where the indigenous cultural heritage was based on the oral tradition, the Europeans completely wrote them off as uncivilized, sub-human kind of people. However in the Indian sub-continent they could not dismiss the indigenous culture that easily. This was a land of economic prosperity. India had a long tradition of language and literature, philosophy, culture, architecture, medicine, science and various other aspects of civilization, considered by the west as true marks of progress. However the Indians were neither white, nor Christian and they valued spirituality and emotionality over cold logicality. The Europeans could not accept the fact that a group of people belonging to a completely different race, culture, religion and belief could achieve even more advancement than the Europeans themselves. They portrayed the Indians as a race once placed at the pinnacle of civilization but now devoid of all past glory and in urgent need for help from the western civilization to regain their lost achievements. The Europeans were represented as the "self" and the Indians or any other colonized race was suitable examples of the "other".

Bhabha, an Indian theorist and critic, has a very interesting take on the relationship between the dominant "self" and the marginalized "other". It is true that the prolonged colonial regime has imposed a tendency of mimicry in the minds of the native people. However this mimicry, replete with "… its slippage, its excess, its difference" (Bhabha, 122), is not an effort to harmonize with the colonizers. It is in fact a way of registering the difference by the colonized "other".

After World War-II, the British before retreating from India divided the sub-continent into various smaller nations based on the religion of the people so that the region will remain divided and in a state of perpetual tension and treachery. Now we have Pakistan, a Muslim nation and India, a secular nation with majority of the people being Hindu. The region experienced another partition when around 1970 East Pakistan demanded freedom from West Pakistan and became the independent nation of Bangladesh with India's military support. The entire region was submerged in a state of communal riots. The task of the indigenous leaders became immensely difficult because whatever they did was interpreted with a religious undertone. The story of *The Shadow Lines* is set in this background. The shadow lines are the imaginary lines which not only signify the international boundaries between India, Pakistan and Bangladesh but keep the people, originally belonging to a single nation, divided for ever.

In *The Shadow Lines* Ghosh alters the narrative perspective. The narrator in this text is an Indian whose name is not revealed to the readers. Ghosh is willfully presenting a non-rigid narrator so that he may incorporate the points of view of different individuals and even groups of people in his narration. This shows that there is no fixed definition of either the "self" or the "other". These are two interdependent concepts. The "self" requires the "other" to define itself. Moreover Ghosh also shows us that it is not only the "self" or the centre which influences the circumference but the periphery also manipulates the centre.

In this novel the character of 'thamma', the grandmother of the narrator, constantly depends upon the "other" for her own existence. She describes the "other" as her enemy. At different stages of her life this "other" or the enemy keeps on changing. As a young girl, she lived in her ancestral house at Dhaka, in a joint family set up. However with the death of her grandfather, the ancestral house was divided into two equal parts. Her uncle's family became the first "other" in her life. Thus the "self" was divided into two parts and they became each others enemy. This is true for India as well. The mother nation is divided into several independent countries which exist in a state of continuous feud with each other.

Later on the thamma considered the British colonial masters to be the enemy or the "other" and was thoroughly inspired to take an active part in India's struggle for freedom. Once the nation attained its political independence, thousands of people suddenly became homeless. Their ancestral land now belonged to a foreign country and in the newly formed nation of India or Pakistan they were refugees. The thamma did not succumb to this calamity. She fought back and established her position in the new homeland. She started from the refugee camp and went on to become the head mistress of a girls' school. The grandmother was able to curve a niche for herself in the male dominated world. All this while the western colonizers were an absent enemy for her. It was after East-Pakistan had become the independent Muslim nation of Bangladesh that she found a new name for her enemy — the Muslims of Bangladesh. Her idea was validated by the incident of cruel killing of her old uncle, the rickshaw-puller and Tridib in the streets of Dhaka as a result of communal disharmony. Incidentally, in this later stage of her life she no longer considered the old uncle to be an enemy any more.

Towards the end of the novel the thamma considers the younger generation to be the "other" because they do not always accept the orthodox lifestyle but are ready to create their own ways of living. The thamma's entire existence is based on a series of negation of the various enemies or "others" at different stages of her life.

The other important women characters in *The Shadow Lines* include Illa and May Price. Illa is a representative of the modern woman in search of a new identity in this newly emerging world. She feels that the traditional Indian identity is too limiting for her. She wishes to be liberated from all kinds of past history which form a vital part of her national, cultural heritage. However her problem lies in the fact that although she is certain about what she resents, she is uncertain about what makes her truly happy. Illa lacks the creative eye. One must have the vision of an artist if one has set out to create a new definition of one's own self.

May Price is an English woman who represents an ideal "other" for any Indian. It is interesting how May becomes Tridib's love interest in this novel. She alters her position from being the "other" to a position very close to the heart of the self. Judith Butler in her book *Gender Trouble: Feminism and Subversion of Identity* has discussed the views of Freud

…the identification with the lost loves characteristic of melancholia becomes the precondition for the work of mourning…Strictly speaking, the giving up of the object is not a negation of the catharsis, but its internalization and, hence, perversion. (Butler79)

This is how even an ex-colonizer may share the Post-colonial space with the erstwhile colonized people. May exemplifies the actual meaning of world brotherhood through her compassion and silent suffering. She is the person who reveals the truth of Tridib's ultimate sacrifice.

The nameless narrator in *The Shadow Lines* and Nick Price are described as each others mirror images. We form our perception about the narrator from his vertically inverted image in the form of Nick Price. By presenting them as each others mirror images Ghosh is further blurring the distinction between the "self" and the "other". The "self" is not a fixed idea. It continuously evolves into a new identity with its interactions with other characters. In this novel although there is a first person narrator, but he too experiences the moment of epiphany along with the readers.

Salman Rushdie in his iconic novel *Midnight's Children* has presented similar character pairs. Saleem Sinai and Shiva are actually interchanged at birth. We are intrigued by the question what would have happened if they weren't interchanged; would Saleem ever be as violent and jealous as Shiva and could the latter ever be a kind and compassionate man. The text is an open-ended one which welcomes various interpretations on the part of the readers.

In *The Shadow Lines* the narration does not have a simple linear progression. Different historical events from the past and present seem to intersect each other at various levels. Ghosh is a master craftsman of personalizing a public incident. He creates ambivalence between history or fact and fiction. The readers are allowed to interpret the novel in their own way. The difference between the author and the reader gets blurred too. The author no longer remains the absolute authority over the narrative of the novel. He has to share his privileged position with his readers and appreciate their participation in writing the text with each reading. The alteration of the "self" and the "other" takes place at various levels.

In *The Hungary Tides* by Ghosh we find different character pairs. They at times work as each others foil and at times are presented as mirror images of one another. Piya and Moyna are apparently two totally different characters.

Piya is a highly qualified marine biologist brought up in America while Moyna is a nurse, a wife and a mother belonging to the Sundarbans. However both of them are truly progressive women. Their passion for their work surpasses every other commitment. They belong to different strata of the society but are representatives of the modern women who are all alone in their quest for a life meaningful to themselves and not predetermined by the society for them.

The most striking contrast is perhaps between the grandmother and Illa on one hand and Tridib on the other in *The Shadow Lines*. The obvious question is whether Tridib is manly enough to be a hero and are the women characters properly portrayed. Tridib is a creative soul. His heroism lies in his ability to imagine a world devoid of the man-made boundaries and to sacrifice his life to uphold his belief. The women are not overtly emotional or imaginative. They make mistakes in choosing the true hero of their life, but are well accomplished in the outer world, the so called male domain. The anatomy of a character does not determine their personality.

There are not only character pairs in this novel, but also cities which appear to be each others mirror images. Calcutta and Dhaka of the 1970s and London and Berlin of 1936 are two such city pairs. During World War-II London and Berlin were the head quarters of the enemy camps, but the life of the common people in these two cities was devastated in the same manner. The day the British authority in London captured the German woman as an emissary of the enemy and May's uncle as a traitor for sheltering her, marks a pitiful predicament in the history of humanity. In a similar manner the attack on the rickshaw-puller and the old uncle of the thamma in the streets of Dhaka is an example of mindless communal hatred.

Tridib actually motivates us to free ourselves of our prejudices. He had taught the narrator the Red Indian way of salutation and a hand shake. He is inspiring a young person to accept the culture of a race markedly different from us. Tridib respects people belonging to separate communities and following different religion. He does not see them as our enemies. Tridib believes that there is no need for any hatred or violence between different communities. We should not be afraid of the differences. Instead accept these variations as uniqueness of the human race and appreciate them. The shadow lines which exist to mark the national borders also segregate human beings on the basis of their race, colour, religion, and various man-made issues. Between Illa and Tridib, the latter is much more global than the former. Ghosh feels that the term 'global- citizenship' should not represent a situation where the individual is unable to adjust to any nation. Illa has traveled all over the world, but there is not a single place which she considers to be her home. She has renounced her Indian identity for a modern and liberal one, but is still a misfit in London. She is a totally rootless person. Tridib on the other hand nurtures a heart large enough to accommodate people belonging to any race or religion. He will be perfectly comfortable in any company at any corner of the world.

Thus Ghosh is trying to point out that the nature of a person cannot be predetermined by any kind of norm or principle. Similarly the differences between the east and the west, the 'self' and the 'other', the protagonist and the antagonist, history and fiction or even that between a man and a woman are not water-tight compartments. We need to overcome these boundaries first in our mind and only then can we really appreciate the essence of such terms as global-citizenship or world brotherhood.

Bibliography

Primary Source

Ghosh, Amitav. *The Shadow Lines*. New Delhi: Permanent Black, 1998.

Ghosh, Amitav. *The Hungry Tide*. 2004. India: HarperCollins Publishers 2005.

Rushdie, Salman. *Midnight's Children*.1981. London: Vintage 2006.

Secondary Source

Ahmad, Aijaz. *In Theory*. 1992. Delhi: Oxford University Press, 1994.

Bhabha, Homi K. *The Location of Culture*. 1994. New York: Routledge Classics, 2004.

Butler, Judith. *Gender Trouble: Feminism and the Subversion of Identity*. 1990. London and New York: Routledge, 1999.

Ghosh, Amitav. *In an Antique Land*. New Delhi: Permanent Black, 1992.

Loomba, Ania. *Colonialism/Postcolonialism*. London and New York: Routledge, 1998.

Zahar, Renate. *Frantz Fanon: Colonialism and Alienation Concerning Frantz Fanon's Political Theory*. Translated by Willfried F. Feuser. 1969. New York and London: Monthly Review Press, 1974.

Chapter 10

CULTURAL CONTRADICTION AND SIGN OF RACIAL ARROGANCE: A CASE STUDY OF THE 19ᵀᴴ CENTURY'S COLONIAL RULE IN INDIA

FIROJ HIGH SARWAR
Assistant Professor, Dept. of History,
Murshidabad Adarsha Mahavidyalayay
(Affiliated to University of Kalyani),
Murshidabad, W. B.

INTRODUCTION:

The commencement of modern nations of race coincided with the European colonization of the world.[8] Consequently, the subject of contact, conflict, and the oppression of people from the "other world" with different physical features and different cultures led to an ongoing debate about the human character of the 'other'.[9] European analysts of the 18ᵗʰ and 19ᵗʰ century used their own civilization as a standard by which other civilization would be judged. Their ethnocentric perspective on the world sometimes led them to

---

[8]   Race, according to New Oxford Dictionary of English, is "Each of the major division of human kind, having distinct physical characteristic, a group of people sharing the same culture, history, language, etc. See the New Oxford Dictionary of English Second Edition 2003. Cambridge International Dictionary of English says that" race is group especially of people with particular similar physical characteristics, who are considered as belonging to the same type are the fact of belonging to a particular such group". See: Cambridge International Dictionary of English 2000.

[9]   Here 'other' and 'other world' meant for the nations other than Europe, where British were able to spread out the umbrella of colonization.

conclude that the "other" was not fully human. Even most of the times, they distinguished their "civilized" world from that of the "barbarians".[10] Comte Arthur de Gobineau's work *The Inequality of Human Race* (1915) set the tone for a century of writing on race, where the goal was not only to classify people along racial lines, but to highlight the presumed superiority of some groups and the inferiority of the others.[11] It was the mighty British Empire, which carried on the same perception of superiority and inferiority in her colonies, and acted as a superior race throughout in America, Africa and India.

British Imperial dogmatism and self-assuredness were strengthened by certain ideas and pseudo-scientific theories. The first of these was the idea of race superiority. The belief in race superiority and its relation to imperial domination was nourished by evolutionary hypothesis such as the *Survival of the Fittest*, the *Aryan Master Race*, and the *Social Darwinism*.[12] James Joll, one of the advocates of the theory, believed that the white races are superior to the black or yellow.[13] This was one of the basic assumptions of the confident imperialism of the 19th century.[14] For instance, "what is empire but the predominance of race" said Lord Rosebury, former P.M. of England.[15]

The concept 'racism' has been defined by Michel Bnton, an expert on the subject of racism, as "the doctrine that behavior is determined by stable inherited characteristics deriving from separate racial stocks, having distinctive attributes, and is usually considered to stand to one another in relations of superiority and inferiority" is more applicable here.[16] There was often the tendency among the Europeans to identify themselves as superior, especially because of two things – white Europeans at the time had a relatively well

[10] Kivisto Peter and Croll, Paul R. *Race and Ethnicity: The Basics,* Routledge, New work, 2012, pp.4-5

[11] Ssuch thinking permeated Social Darwinian ideology. Ibid., p.6

[12] The terms "a master race, racial superiority, the special creation, the survival of the fittest, natural selection, existence for survival" etc are coined in by the European to confirm their racial superiority.

[13] Black and Yellow colour indicates the African and Indian people.

[14] Joll, James, *Europe Since 1870*, London, Pelican, 1983, p.104

[15] Shreedharan, E., *Textbook of Historiography: 500 B.C. to A. D. 2000*, Orient Longman, Hyderabad, 2004 p.409

[16] Banton, Michel. *Race Relations,* Tavistock, London, 1967, p.19

developed technology, and the very fact that the Europeans were able to defeat non-Europeans in war showed that in terms of evolution and progress they were fitter to survive than the non-Europeans were to do.[17] That is why due to the scientific advancement and progressive vision, European, since 17th century, kept maintaining their superior hold on the human as well as natural resources of the world.[18] Thus, someway and other, their claim as advanced race is accepted logically. However, in turn, the claim of superiorities affected and undermined the basic human values in 'other world', which questioning their superior character. The western belief of biological inferiority and superiority often produce negative treatment for the backward nations of the East.[19] The twin proposition – "racial discrimination versus civilization mission" of European colonizers became a common colonial scenario in 19th century that has been depicted in many contemporary literatures. For example, E.M. Forster, a notable writer, in his novel *A Passage to India* has explained the philosophy of racialism vividly.[20]

In the case of British India, as a part of self-justification, it was generally believed complacently that Indians were unfit for self government, and British rule was the best for them. In the course of action as the British developed a notion of superiority (superiority of race, civilization and power), they thought they had right and authority to govern the others for the sake of righteousness of the world. J. A. Hobson justifies the British racial superiority as "it is desirable that the earth should be peopled, governed, and developed, as far as possible, by the races which can do this work best, i.e. by the races of highest

---

[17] Sreedharan, E. Op. cit., p.409

[18] Due to the European renaissance and their subsequent developments in science and culture, European nations became first modernised race in the world, and therefore they dominate all recent developments in the world.

[19] Driedger, Leo. *Race and Ethinicity, Finding Identities and Equalities,* Oxford University Press, Canada, Second Edition, 2003, p.216

[20] His novel is a reflection of his insight in which various occasions, characters, and incidents have been discussed where biased and prejudiced attitude of European communities in India is exposed. See the novel of Forster, E.M. *A Passage to India.* London: Penguin Books.2005; Rudeness, Race, Racism and Racialism in E.M. Forster's "A Passage to India" Gulzar Jalal Yousafzai & Qabil Khan, Dialogue,_January to March, 2011, pp.75-92

'social efficiency."[21] Vincent Smith marked that endemic political chaos was the normal political condition of India. Therefore, it made incapable Indians to unite and rule themselves and made the permanence of British rule absolutely necessary.[22] As earlier sir Thomas Smith argued, God had given the English responsibility to 'inhabit and reform' this 'barbarous' nation.[23] As the product of a conception of civilization whose differing levels secured a place of the English as its apex? The English took this rational for the subjugation of foreign peoples from Ireland to America, and then to India and to Africa.[24]

Since the conquest of Ireland in the sixteenth century, the English gradually emerged as the "New Romans", stimulated with the ideas of civilizing backward peoples across the world.[25]In a post-Enlightenment intellectual environment the British started defining themselves as modern or civilised vis-a-vis the Orientals. The feeling of the 'chosen race' gradually developed in the minds of Englishmen that coincides the 'task of civilizing mission', which Ralph Fox termed as 'the psychology of the civilized bandit'.[26] This rationalized their imperial vision during the 19th century.[27] Late in the same century, British began to belief that they had discovered the secret of progress through the employment of reason. Simultaneously they put Eastern World, particularly India, in the category of static or semi-barbaric cultures. Based on this conviction, different views were formed as to how Indian society

---

[21]   Hobson, J. A, *Imperialism: A Study*. Cosimo, Inc., 2005, pg.154

[22]   Sreedharan, E., Op. cit., p.427

[23]   Metcalf, Thomas R., *Ideologies of the Raj; The new Cambridge history of India*, Vol. 3, Cambridge: Cambridge University Press, 1994, p.2

[24]   Canny, Nicholas, 'The Ideology of English Colonization: From Ireland to America,' *William and Mary Quarterly*, 3rd Series, vol. 30, 1973, pp.575-98.

[25]   Metcalf, T. R., Op. cit, p.3

[26]   Fox, Ralph. *The Colonial Policy of British Imperialism (introduction by Ian Talbot)*, Oxford University Press, UK, 2008, p.33

[27]   The scope of the Enlightenment project was de constructive in nature. Enlightenment intellectual believed it was necessary to purge oneself of superstitious beliefs before one can being to formulate new ideologies. But these new ideologies were inherently unsystematic. It was the belief in the power of knowledge, rather than adherence to any particular belief system, that resulted from the Enlightenment. See: Vandana Joshi, (edit.) *Social Movements and Cultural Currents, 1789-1945*, Orient Blackswan Pvt. Ltd., New Delhi, 2010, p.278

was to be redeemed and reclaim for civilization. In England, David Brown, Henry Martyn, George Udney, John Shore and Charles Grant represented the evangelical viewpoint, and all believed that Britain had a civilizing mission in India to be achieved by replacing the values of Indian society by British ones.[28] In this context, Claudious Buchanan held that "only through Christianity could true civilization be imparted."[29] Their ideas got worldwide publication through a treatise of C. Grant, namely *"Observation on the State of Society among the Asiatic subjects of Great Britain.*[30] In this Grant argued for the application of Christianity and western education to change what he thought was a 'hideous state of Indian society'.[31] Accordingly, by the Act of 1813 and Act of 1835 the suggestions were implemented through allowing Missionaries to be settled in Indian and by introducing Western Education.

However, after 1857, the whole scenario got changed, and the Englishmen began to follow rather a sharp, even sometime arrogant attitude towards Indian. The writers like James Fitzjames Stephen, whose writing in the 1880s, contended that empire had to be absolute because "it's great and characteristic task is that of imposing on Indian ways of life and modes of thought which the population regards without sympathy, though they are essential to its personal well-being and to the credit of its rulers." The above influences issued in a theory which asserted that the strongest always ruled.[32] Fitzjames, also popularly known as Social Darwinist, not only emphasized Indian's difference, but also asserted Indian's inferiority. Such ideas in the nineteenth century

---

[28] Angelicalism: a protestant Christian movement in England of 18th century, which in contrast to the Orthodox Church emphasized on personal experiences, individual reading of gospel rather than the tradition of established Church. See: Ingleby, J.C. *Missionaries, Education and India: Issues in Protestant Missionary Education in the Long Nineteenth Century,* ISPCK, Delhi, 2000, pp. 5 and 2

[29] See: Buchanan, C. A. *Sermon Preached* at the *New Church of Calcutta,* Calcutta, 1800

[30] To Grant Indian civilization was barbaric because its religion was degrading, which was both dangerous and a violation of the Christian spirit even to tolerate such a culture. Embree, A., *Charles Grant and British Rule in India,* Colombia University Press, New York, 1962, p. 148

[31] See Mangalwadi, V. *India: The Grand Experiment,* Farnham, 1997 for Grant's influence. See Charles Grant, *Observations on the State of S~ciety among the Asiatic Subjects of Great Britain, particularly with respect to Morals and the Means'D.f improving it, 1792.*

[32] Sreedharan, E. Op. cit., p.412

were further strengthened by the rise of racial sciences in Victorian England, which privileged physical features over languages as the chief markers of racial identity.[33] Hence, the story of invading white Aryans founding the Vedic Civilization through a confrontation with the dark-skinned Indian aborigines was invented. It was a theory, which constructed by "consisting overriding" of evidence and "a considerable amount of text-torturing".[34]

By the end of the century, Rudyard Kipling, in his famous poem, The White Man's Burden (1899) put this case more comfortably for the European, as he writes:

Take up the White Man's Buden
Send forth the best ye breed
....
To serve your captive's need;
....
Your new-caught sullen peoples,
Half-devil and half-child.[35]

The imperialist argument based on the White's man's special right to rule was given a moral and humanitarian cover. The Christian missionary was altruistically eager to save heathen souls from the certainty of hell. In the very beginning, William Warburton, bishop of Gloucester, in a sermon, preached in 1766, stated that no lasting conversion of Indians could be expected until "these barbarians" have been "taught the civil arts of life".[36] According to the European bureaucrats - the non-whites were totally incapable of undertaking tasks of leadership and ruling themselves. Therefore, tutelage of the civilized

33   Bandyopadhayay, Sekhar. *From Plessey to Partition – A history of Modern India,* Orient Longman, New Delhi, 2004, p.73

34   Metcalf, T. R. Op. cit., p. 208

35   See the Rudyard Kipling's poem, *The White Man's Burden,* 1899. The responsibility of governing India had been placed by the inscrutable providence upon the shoulders of the British race. Most English officers and man, serving in India, followed the ideals laid down by the poet Kipling in his poem *'The White Man's Burden'* (1899). Having overcome initial inertia, they settle in their respective duties according to their education and training.

36   Marshall, P.J., *The making and unmaking of Empires,* Oxford University Press, New York, 2005, p.192

over the uncivilized was thought to be a necessary. In the late nineteenth century, a psychology of imperialism developed and found utterance in such prominent men as Thomas Carlyle, Charles Kingsley, Benjamin Disreali and John Ruskin. John Selly's *Expansion of England* (1883), one of the sources of British imperial psychology, was so enormously popular as to get into the reading of most middle and upper class school boys.[37] During those days, British intellectuals used to speak and write of their country's advancement by quoting certain phrase, viz. 'civilizing mission', 'advances of the flag', and 'manifest destiny'. Such were the language, words and phrases, used to cover up Europe's power impulse that overlapped the true nature of mad race - earsplitting for empire and spheres of influence dictated by the capitalist search for raw materials, consuming markets, and fields of investment.

II

SIGN OF RACIAL ARROGANCE:

The British always sought to supplement their control of the Indian Empire through a web of hegemonic practices involving subtle strategies of cultural manipulation. Knowledge of Indian culture, ideology, ethnology, ethnography, anthropology and the geography of India helped the British colonizers to build up a powerful discourse. Though, few English men rather orientalists - Warren Hastings, William Jones and Max Muller studied Indian past, and admired and popularized the greatness of Indian civilization in the world, however, that made nothing pragmatic change in the perception and attitude of the British. Their philological study, which confirmed the kinship between European and Asian race, proved meaningless for the governing body of England.[38] The men like James Mill and Charles Grant discarded the theories of ancestral relation between Indian and British, and depicts Indian as a semi-barbarous and uncivilized nation. The early colonial historiography shaped and reshaped the policies of governances in India. In this context, most prominent work is of J.

---

[37]   Furber 'The theme of imperialism in modern writings on India', in Philips, edited. *Historians of India, Pakistan and Ceylon*, Oxford University Press, London, 1967, p.341

[38]   It was James mill who made an effort to prove the kinship between various western and Asian races through the philological link and citation. See: Trautman, Thomas R. *Aryans and British India*, Yoda Press, New Delhi, Indian Published 2004, pp. 45-50

Mill, whose writings asserted that the laws and the institutions of the Hindus could not have began or existed "under any other than one of the rudest and weakest states of the human mind".[39] He declared that "the people of Europe, even during the feudal ages, were greatly superior to the Hindus". "In truth", writes Mill "the Hindu, like the Eunuch, excels in the qualities of a slave".[40] The tremendous influence of Mill's book *History,* had on British policy towards India, could be seen in the numbers of times it went to press -1818, 20, 26, 40. According to C. H. Philips, 'the book serve as main basis for British thought on the character of Indian Civilization and on the way to govern India.' The book was also established as a textbook at Haileybury collage from 1805-1855, where the Company's civil service recruits were trained, and where a succession of eminent utilitarian or close sympathizers held senior teaching post.[41]

Racial doctrines openly preached the predestined superiority of the whites (British) and the permanent subjugation of the non-whites (Indian) to the white supremacy. As a result the British not only enjoyed numerous exemptions and privileges but also they became so far brutalized as to insult, assault and even murder Indians with impunity. This quite obviously moved self-respecting Indians to challenge the odious alien rule. The early sign of racial discrimination in India could be seen in the Cornwallis code at the end of the 18th century that rapidly crystallized in the 19th century also.[42] The code itself suggested the un-trustworthy character of Indian, and therefore, Indian would not be part of the administration. Incidentally, the process of eliminating Indian from key government post has already been started with the Reza Khan, first Diwan of colonial Bengal. One of the main reasons for arresting Reza Khan in 1772 and for keeping him for confinement without trial for nearly two years was to get rid of the most powerful obstacle to this project of eliminating Indian agents from the administration of Justice.[43]

---

[39] Mjumdar, R. C., Historiography in Modern India, Asia Publishing House, 1970, p.12

[40] Ibid. p. 13

[41] Sreedharan, E. Op. cit., p.404

[42] See: Aspinall, Arthur. <u>Cornwallis in Bengal</u>, Manchester University Press, UK, 1931

[43] Khan, A. M. The Transition in Bnegal, 1756-1775, Cambridge University Press, Cambridge, 1969, p.294 & 341

If we look at the actual functioning of the empire, the statements of racial superiority of the rulers were made rather prominent since the late eighteenth century. It was Governor General Cornwallis who transformed the company's bureaucracy into an "aloof elite", maintaining physical separation from the ruled. British soldiers were forbidden to have sexual relations with Indian women. Moreover, the company's civil servants were discouraged from having Indian mistress, urged to have British wives and thus preserve – "the respect and reverences the natives now have for the English".[44] Any action undermining that respect, Henry Dundas, the president of the Board of Control, had argued as early as 1793 would surely "ruin our Indian empire".[45]

In this sphere, one argument was repeated in a confidential circular issued by Sir Alexander Mackenzie, one of the British chief Commissioners, in 1894. He was more concerned to emphasize the danger to racial prestige. He wrote, who had Asian mistress, not only degrades himself as an English gentleman but lowers the prestige of the English name and largely destroys his own usefulness. He repeated Crosthwaite's warning about the danger of a growing Anglo-Burman class, who would probably bring discredit on the ruling race.[46] He repeated the same in a variety of news paper from Rangoon 'Times' to 'the Times of India'. At a time, he made a proposal to punish those British, who have relations with native women, by stopping their promotion and upgradation.[47] In the same line, the officiating commissioners of Patna feared political danger and suggested to maintain the enough distance between the ruling race and the Indian people.[48]

Such overt statements of physical segregation between the ruler and the ruled as an ideology of empire were quite clear in the very way the human environment of the imperial capital city of Calcutta developed in the eighteenth century. The process worked in an overall setting of dualism, basically a feature of all colonial cities, between the white and the black town.[49] This phenomenon

---

[44] The appeal was made by an official before the parliamentary select committee in 1830

[45] Ballhatchet, Kenneth. 1980, pp. 2-3, 96-97

[46] Ibid., p.147

[47] Ibid., p.147

[48] Ibid., p.124

[49] Sinha, P. *Calcutta and Urban History*, Firma K. L. Mukhopadhyay, Calcutta, 1978, p.7

of dualism reflected on the one hand, the conquerors' concern for defense and security, but one the other, their racial pride exclusivisim. In the early 18th century, this special segregation along racial lines had been less sharply marked, as there was a 'White Town', and a 'Black Town', intersected by a 'Grey Town' or an intermediate zone, dominated by Eurasian or East Indians. The position of the Eurasians, the children of the mixed marriages (popularly known as half-caste) continually went down in the imperial pecking order since 1791. They were debarred from covenanted civil and higher-grade military and marine services. Major-General Jasper Nicolls testified in 1832 that Eurasians should not be given Commissions in the army.[50] The racial polarization of colonial society was now complete. By the early nineteenth century, "the social distance" between the people and the ruling race became an easily discernible reality in Calcutta's urban life.[51]

The British people living in different parts of the country developed exclusive societies for themselves. The British people also lived in places particularly demarcated for the white-skinned people where no 'native' were allowed to enter. In Calcutta, for example, the habitation area was particularly divided into two parts, namely, the 'black' and the 'white' areas, obviously discriminating between the Indians and the Europeans. No Indian was allowed to enter the areas exclusively marked for the white-skinned people. Likewise there were parks, gardens and play grounds where no Indian was allowed to enter. In the case of reservation of compartments in railway and steamers, a line of racial discrimination was completely followed by the British. Even after having enough material capacity, Indian was barred to enter in certain coaches and apartments. In *A Passage to India,* Forster's emphasis on the racial differences received more importance than that on cultural differences. He narrated that at Chandrapore Club all member are white and no Indians are allowed there. In response to Mrs. Moore's insistence Aziz retorts that no Indians are permitted to enjoy the show in the club. Even "windows are barred, lest the servants should see their mem-sahibs acting".[52] In fact, the colonizers never truly intended to interact with the colonized as they believed

---

50 Ballchatchet, Kenneth. 'Race, Sex, and Class under the Raj; Imperial Attitude and policies and their critics, 1793 – 1905', Vikash Publishing House, New Delhi, 1979, p.98

51 Ballhatchet 1980, pp. vii, 97 and passim

52 Foster, E. M. Op. cit., p.25

only in domination and submission. British were never in mode of any kind of civil relation with Indian, and this temperament turns the colonizer into a master, a slave driver, a prison guard, above all, a god.[53] All this made the Indians conscious about the national humiliation and accelerated the pace of disillusionment with the foreign rule which eventually turned them anti-British.

Imbued with an ethnocentric sense of superiority, British intellectuals, including Christian missionaries, spearheaded a movement that sought to bring Western intellectual and technological innovations to Indians. The immediate consequence of this sense of superiority was to open India to more aggressive missionary and proselytizing activity in India. Initially British in India were conscious that any attempt to convert Indians to Christianity promised to subvert the very foundations of civil peace by offending the most deeply entrenched religious prejudices.[54] However, by the passing of time, government changed the notion, and allowed the European missionaries to settle in India in 1813. The proselytizing mission of missionaries, and their overemphasize on Christian light as a means of eliminating Indian backwardness, simply suggest that British was also eager to maintain superior hold of Christianity over India religions.

In the sphere of judicial system, we also can see a sharp line of discrimination on the basis of race. Jaffe James argued that the race played perhaps the single most prominent role in the construction of the British judicial system in India.[55] The entire judicial reform of Cornwallis spoke of one thing – a total exclusion of Indian from the whole system, which became less ambiguous in its authoritarian and racially superior tone.[56] The concept of equality before law often was not applied to the Europeans. If there was greater movement towards equality in civil justice system, racial privilege for the rulers remained in place

---

[53]   Elham Hossain, Muhammed. 'The Colonial Encounter in a Passage to India', *ASA University Review*, Vol. 6 No. 1, January–June, 2012

[54]   Edward, Michael 'British India, 1772-1947; A Survey of the nature and effects of alien rule', Sidwick and Jackson publisher, London, 1967, p.54

[55]   Jaffe, James. (Review) 'Colonial Justice in British India: White Violence and the Rule of Law" by Kolsky, Elizabeth, H-Law, *H-Net Reviews*. November, 2010

[56]   Bandyopadhayay, S. Op. cit., p.99

in various forms in the criminal courts.[57] In 1826, a Jury act was passed which introduced religious discrimination in the law courts. Under this Hindus and Muslims could be tried by European or Indian Christians, but no Christians whether European or Indian, could be tried by Hindu or Muslim jurors.[58] The Ilbert Bill storm was the most extreme but by no means isolated expression of such white racism. It shows quite apparent of British attitude towards Indian when British questioned – why should we be judged by a nigroo (India)? In 1878, for example, the appointment of Mathusamy Iyer as High Court judge in Madras was opposed by the Madras Mail (organ of White businessmen) on the ground that 'native officials should not draw the same scale of pay as Europeans in similar circumstances".[59] In both case of the debates, surrounding the 1861 Code of Criminal Procedure and the Indian Evidence Act of 1872, the resistance of the white "non-official" community played a significant role in deflecting government attempts to introduce "an equal and uniform law of jurisdiction" for both Britons and Indians.[60] Instead, British subjects eventually were guaranteed the privilege of appearing before European judges and thus were permitted to avoid having their cases heard in the common local courts which were staffed by Indian judges and magistrates.

In the name of equality of Law, Englishmen usually received preferential treatment in the courts. Nonetheless, It may be argued that while many of the cases discussed undoubtedly substantiated the author's contention that somewhat misleading and perhaps inaccurate to employ such terms as "lawlessness" and "practical impunity from prosecution and punishment" to describe the situation.[61] Horrific and brutal as the cases described here are, their history is nonetheless recovered from British court records and thus bear witness to the active prosecution and often conviction of white settlers.

---

[57] Singha, Radhika. A Despotisim of Law: Crime and Justice in Early Colonial India, Oxford University Press, Delhi, 1998, p. 30, 289-93

[58] See a comprehensive history of Modern India in internet, p.186

[59] Sutharalingam, R. *Politics and Nationalist Awakening in South India, 1852-91*, University of Arizona, Tucson, 1974, pp.152-2

[60] Kolsky, Elizabeth. Colonial Justice in British India: White Violence and the Rule of Law. Cambridge University Press, Cambridge, 2010, p. 10

[61] Ibid., p. 35

Without doubt, the British courts in India were often inept and even more often biased and unfair.[62]

In the sphere of government job, both Company's and Crown's rule followed a segregation policy very sophisticatedly. The British followed a policy of exclusion of Indian from very key and senior posts in the military and administrative cadre as much as possible. From the post of civil service, Indians were carefully excluded, as no position worth an annual salary of 500 pound or more could be held by them.[63] The creation of covenanted and non-covenanted post and the reservation of covenanted post for English under British rule was fully a discriminating recruitment policy based on race. The eligible claims of the India for various posts used to undermine by a racial stereotype, which implied the deficiency in character and morals of the native.[64] Gradually, the circumstances got certain adjustment. It was after 1858, the civil service Commission henceforth started to recruit civil servants through an examination held annually in England. With the growing percentage of educated Indian, the aspirants of the examination for the post of civil servant kept increasing. However, the British government consecutively abridged the age limit in such a manner, in which, it became impossible for the Indian to seat in the exam.[65] Furthermore, the introduction of the 'Statutory Civil Service" in 1870 declared that the Indian of ability and merit could be nominated to a few position hitherto reserved for the European covenanted post. British objected the very idea of introducing the principle of election in India and obstructed the proposed Indianization of the civil service on the basis of a "mythical rational" of "inefficiency" that was used to legitimize their own monopoly of power.[66] Mr. Elgin argued in a letter to Roseberry in July 1895 that "we could only govern by maintaining the fact that we are the dominant

---

[62]   For detail see: Mallampalli, Chandra. *Race, Religion, and Law in Colonial India, Trials of an Interracial Family,* Cambridge University Press, Cambridge, 2011

[63]   Bandyopadhayay, Op. cit., p.110

[64]   Ballchatchet, Kenneth. Op. cit., p.99

[65]   The maximum permissible age was gradually reduced from 23 (in 1859) to 22 (in 1860), to 21 (in 1866), and to 19 (in 1878). See: Ahir, Rajiv, *A brief History of Modern India,* Spectrum Bokks (p) LTD, New Delhi, 2008, p.305

[66]   Spangenberg, B. *British Bureaucracy in Indian: Status, policy and the ICS, in the late Nineteenth century,* South Asia Books, Colombia, 1976, p.347 and passim

race - though Indians in services should be encouraged, there is a point at which we must reserve the control to ourselves, if we are to remain at all".[67]

Indians not only suffered discrimination in the matter of recruitment, but also in the field of promotion and pension. The salary, pension and allowance were always paid less to the Indian whereas the European received much more for the same post.[68] It was very ostensible in the military department. The new service act of 1856, worsened the matter bitterly among the Indian sepoy. It disqualified Indian soldiers from the distance allowance facility. For the case of promotion, the matter was more pathetic. For instance, in the beginning of the 19th century, an Indian police could only become the Habildar (head of the local thana). The police commission of 1902 provided for the appointment of educated Indians to the positions of officers in the police force, but they stopped in rank where the European officers began".[69] Thus distrustful of the Indian subordinates and subservient to the civil authorities, the Indian police system was tellingly reflective of its colonial nature.

In addition to these, often Indian officials were treated by the British very badly and received quite uneasy reaction from the government. Most of the time, Englishmen were not obeying and followed the orders and policies, instructed or prepared by Indian officials. For instance, one evening in September 1860, a drunken English officer gave much trouble and a guard had to be called out to deal with him. As matter of fact, he was not obeyed that Indian policemen and on very next day he was rebuked by the officers.[70] For the less fortunate, racism took cruder forms of kicks and blows and shooting 'accident' as the 'sahib' (Englishmen) disciplined his *punkha* coolie or bagged a native by mistake while out on *shikar* (hunting). No less than 81 shooting 'accidents' were recorded in the years between 1880 and 1900. White dominated courts regularly awarded ridiculous light sentence for such incidents.[71] On the other hand, white soldiers

---

[67]   Sarkar, Sumit, *Modern Indian, 1885-1947*, Macmillan, New Delhi, 1983 p.23

[68]   Bandyopadhyay, S. Op. cit., p.171

[69]   Bayley, D. H. *The Police and Political Development in Indian*, Princeton University Press, Princeton, NJ, 1969, p.49

[70]   Ballchatchet, Kenneth. Op. cit., p.124

[71]   The treatment of coolies on Assam tea plantations figured prominently in the work of the Indian Association in the late 1880s. Sarkar, Sumit, Op. cit., p.22

were also frequently offended brutally and harassed (some time it extends up to the rape and murder) the indigenous people without any relevant cause, and for the same, they were not liable to the authority.

By the beginning of the 20th century, a new form of racism took shape in Indian, based on economy. Historian Amiya Bagchi stated that the more crucial form of racism could be found in the economic sphere of life. It has been found that the colour played an important role in preserving the unity of white businessmen in India against possible Indian competitors. The various white associations, namely Chambers of Commerce, The Trade Association and the other organization in the jute and tea plantation reveal that "European traders and businessmen were great believers in reasonable compromise and mutual accommodation among themselves; however, much they might believe in the virtues of competition for others".[72]

III

ASSESSMENT OF CIVILIZING MISSION:

The response of the colonized to the colonizers, hegemony paves the ground of the relationship between them. Very often natives absorb colonial hegemony automatically through a mistaken belief that it was of their own. They had surrendered themselves to foreign rule through cultural intersection. Living close to the outsiders and under their influence for a long time both inadvertently and naturally, made natives absorb their thoughts, ceremonies, rituals, dress pattern and many other phenomena. The moment when Indian reached at the verge of consciousness, they started experiencing the discrimination on the basis of race. As consequence, natives at one point began to think of themselves from the point of view of outsiders.[73] Henry Cotton in his *India in Transition* freely admitted that the most active manifestations of English opinion have often been actuated by race animosity.[74] Sugata Bose and Ayesha Jalal furnished a different kind of view. To them, the early phase

---

[72] Bagchi, Amiya Kumar. *Private Investment in India 1900-1939*, Cambridge University Press, 2007, (reprint) p.170

[73] Elham Hossain, Muhammed. Op. cit.

[74] Cotton, Henry, *India in Transition*, B. R. Publishing Corp, Delhi, 1885, p.185

of British rule in India may have a period of aggression and economic plunder but it was not one of heavy handed social intervention by conquerors imbued with a sense of racial superiority.[75] However, it cannot be denied that the perceptions and attitudes of Englishmen like James Mill and other Evangelical leaders precisely indicate the racial attitudes. Whatsoever, we can see the reality of distinction from the discussion.

As Wilberforce has explained in a way that "let us endeavor to strike our roots into the Indian soil, by the gradual introduction and establishment of our own principles and opinions; of our laws, institutions, and manners; above all as the source of every other improvement, of our religion, and consequently of our morals' for the happiness of Indian".[76] By the 1850s, the British in India had virtually institutionalized their contempt for things Indian. Their sense of duty was fully supported by a militant Christianity which can be seen at its most aggressive in the careers of those English men who had been ruled different parts of India. By the 1857 it was generally felt by the British officials in India that Indians were a pretty evil lot and that it was Britain's duty to civilize and Christianize them. The non official's community was indifferent to Indians, and to Britain's duty to improve them. For its part, it was more concerned over the growth of an educated Indian middle class, which already began to make demands. The non-officials community was anxious to safeguard its own interest against attack from any direction, including the government in London.[77]

In 1859 the late sir Alfred Lyall, then young civil officers, had written from the seclusion of an up-country district: "I am always thinking of the probable future of our Empire, and trying to conceive it possible to civilize and convert an enormous nation by establishing schools and missionary societies. Also having civilized them and taught them the advantage of liberty and the use of European sciences, how are we to keep them under us and persuade them that it is for their good that we hold all the high offices of the Government?"[78] But

[75] See: Bose, Sugata & Jalal, Ayesha. *Modern South Asia; History, Culture, Political Economy*, (2nd edit), Rutledge Publisher, New York & London, 2004., p.61

[76] Edward, Michael, Op. cit., p.54

[77] Ibid., p.36

[78] Lovett, Verney, *A history of Indian nationalist movement*, Low Price Publication,

at a time, it had been argued that it was Europe's mission to civilize India and the European hold it as a trust until Indians proved themselves competent for self-rule. Surprisingly, the hundred year's British civilizing mission could not come up with enough civilized Indian who could hold the responsibility of governance. If we assess the entire development then the causes might be two types. One, either the so called civilizing mission was at very slow in nature or it was a completely hollow proposal.

The facts itself talks that from the very beginning of the 19ᵗʰ century, Evangelicals started its crusade against Indian barbarism and advocated the permanence of British rule with a mission to changing the very "Nature of Hindustan". This could be effectively changed through the dissemination of Christian light.[79] This means they were not in mood for any real development for Indians, they just wanted to maximize the size of Christian population in the World by altering the long cherished self Identity of natives. So where is the real value of civilizing mission which was meant for human kind, not for any self-seeking purpose? Furthermore, the nature and meaning of civilizing mission varied from time to time and from man to man. Some wanted to confer the lesson of civic life to the Indian as they could became the consumer of British goods and institution. Others were in mode to convert the Indian culture into an alien Christian culture.

On the eve of and after the revolt of 1857, when British received huge political frustration, they suddenly changed their attitude and perceptions and molded the principle of Civilizing mission accordingly. It was Victorian liberalism in post-1857 India that certainly made paternalism that dominant ideology of the Raj. After the revolt it was argued that the Indians could never be trained to become like Englishmen.[80] Thus, it was a kind of oscillating mentality of British Raj. Therefore, when the Indian political situation became favorable they surcharged the principle of ethical rule and civilization mission for the Indian to appease the compensation of colonial exploitations through educational and social reform scheme. Whereas, the political situation became unfavorable or Empire suffered from any kind of native resistance, British left

---

Delhi, 1920, pp.20-21

[79]   Bandyopadhaya, Op. cit., p.71

[80]   Ibid., p.73

out the mission of civilization and followed the policy of racial segregation and condemnation.[81] If really British, as a part of their ordain duty, were determined to civilize India then they would not drain Indian wealth to England, rather they would utilize it for the cause and real development of Indian people and make Indian self-dependent Nation.

To draw a conclusion of above discussion we may share the ideas of Sumit Sarkar as who recognized the British in India were quite conscious of being a master race.[82] He further stated that, Europeans constructed a sense of self for themselves, and they had of necessary to create a notion of an "other' beyond the seas.[83] To describe oneself as "enlightenment" meant that someone else had to be shown as savage or vicious. Such projection was an integral part of the enlightenment project. As the British endeavored to define themselves as 'British', and thus 'not Indian', they had to make of Indian whatever they chose not to make of themselves.[84] British usually defined their own identity as a nation in opposition to the World outside. And thus, under the influence of the ideals of Enlightenment, announced their own pre-eminence as a 'modern' and 'civilized' people. The East was always described through the forms of western iconography, and had labeled the British Empire as greater Britain.[85] According to Grant Event School:[86]

---

[81] Condemnation basically imply on the character and civilization of the Indian. They used to alleged that Indian were semi barbaric and uncivilized. To them Indian have no civic society and manner, and they loved to live in chaotic environment.

[82] Sarkar, Sumit, Op. cit.,p.22

[83] Apart from the imperialist historian it is generally believed that the feeling of racial superiority was a vital ingredient of imperialism. Therefore it ran in the English blood and laughed all political difference to scorn. Englishman tried to justify through explaining their superiority of race, first on Religion and then science, were popular in the 19th century Britain, and both were pressed into service. The contemporary command decided that Britain must rule the waves, the sands and the plains because the English race was the elect of all according to scientific principles. But the spiritual energy of the English race, partially suppressed by the 18th century emphasis on reason. See: Aziz, K. K., *The British in India; A study in Imperialism*, Indian Institute of Applied Political Research, New Delhi, 1988, pp.18,.35-38

[84] Metcalf, T. R., Op. cit., p.6

[85] Ibid., pp. 4-5

[86] This school saw history more as the natural evolution of the progress of the British Civilization and liberty, and this historian examined events of consequences in this

"The British State and its expansion throughout the globe was a benign or even positive phenomenon for the British people and the rest of the world. With Britain, came progress, civilization, and most importantly the political freedom inherits in the sublime British constitution."[87]

While 'racism' is a negative concept, based on the belief that some races are superior and others are inferior, the concept of equality is an attempt at ranking people more objectively on the basis of opportunities to compete in the social, economic and political spheres of our society. It is assumed that humans have roughly the same abilities, but for many reasons they do not all have the same opportunity to fulfill their potential' – said by Leo.[88] The white European industrial dominance over colored African and Asians as preindustrial suppliers and servants was a common pattern that emerged around the world.[89] Most countries of European created colonies that supplied raw materials for the European industrial revolution which always benefited white.[90] Ideologies of racism may develop where, in a system of dominance that socializes and communicates differences based on biological features, minorities became less valued.[91]

According to Weber, cultural differences in clothing styles and grooming, food and eating habits, and the division of labour between sexes can all be the focus of a consciousness of a kind that can became either shared characteristics of identity or barriers between groups.[92] Leo Driedger says "Groups who have

---

light. The intellectual harbinger of this school was the 19th century Regis Professor J. R. Seeley argued in a series of lectures published as the "Expansion of England" in 1883, that the British Empire was a gradual and pacific phenomenon which spread commercial and political liberties around the globe. One of the additional innovations made in the "Heroic" school was the idea of the "organic growth" of the British Empire.

[87] Wormell, Deborah & Seeley, John, *The Uses of History*, Cambridge: Cambridge University Press, 1980, p.131

[88] Driedger, Leo. Op. cit., p.216

[89] Berger, Thomas R. *Fragile Freedoms: human rights and dissent in Canada*, Clarke, Irwin, Toronto, 1981, pp.111-136

[90] More sophisticatedly it has been justified that colonizer has right to extract the resources from the colonized nation in exchange of light of civilization

[91] Driedger, Leo. Op. cit., p.4

[92] Guenther Roth and Claus Wittich (eds) *Economy and Society*, Vols. 1 & 2, University

achieved control over certain resources (e. g., capital or institutional authority) will protect their gains and attempt to extend them. They are likely to try to deny access to power and control to other groups to bring about change. The relatively disadvantaged many indeed pressure for change in the economic or political situation, but their chances of success depend on the very factors which brought about inequality. As a result, improvements in the situation of disadvantaged minorities are frequently very slow".[93] Racial prejudices lie at the root of all antagonistic forces against passage of Europe and India. It produces some corrosive differences that cannot be purged. Thus, Stephen Ignatius Hemenway says: "racial differences can never be eliminated, but they can be minimized or overlooked as inconsequential".[94]

of California press, Berkeley, CA, 1978, p. 387

[93] Driedger, Leo. Op. cit., p.19

[94] Hemenway, Stephen Ignatius, *The Novel of India: The Anglo-Indian novel*, Writers Workshop, University of Michigan, 1975, p. 98

Chapter 11

"The Race Industry is a Growth Industry": Tracing the Politics of Representation in Select Poems of British Dub Poet Benjamin Zephaniah

Ayon Halder
Research Scholar
Department of English, University of Kalyani

Colonial encounter results in dismissing the native culture which is reasserted by the nationalist discourse that traces the existence of a culture antithesis to the colonial one. The resistance against stereotyping of the heterogeneous culture of native people is exerted by resorting to the categories that have been deployed by the colonizers to validate the starkly visible difference between the two cultures. This is strategically maintained by the people from privileged class that justifies the colonial invasion by pointing out the sheer irrationalism and barbarity characterizing the intrinsically primitive nature of the racial other. More often than not this is countered by the nationalist writers who take recourse to the dichotomy embedded in Eurocentric literature to rebel against the identification by recognizing the point of departure without denying the distinction made by the colonial masters. The celebration of the primeval form of existence is set against the degraded notion of pigeonholing the colonized entity. Thus the characteristics for which the ethnic people from the colony are severely denigrated are manipulated by the native people who articulate their modes of dealing with life at large by reaffirming the irrationality or primitiveness by inscribing these traits in the nationalist discourse. The colonizers who tend to dislodge the indigenous people from the settler colony or relegate them to the margin are inclined to propagate that these people are devoid of any sense of culture or civilization as they are close to the natural world without a sense of history as compared to the white people who bear the responsibility to drag them out of their debased condition. But the emphasis on the primitive existence on the part of the colonized people while they write back at the Empire speaks volume about the extent of dependence on own ethnic heritage that stands in contrast against the mark of shame labelled on them. This is perceived in the literary movement named as Negritude by the Black French speaking people who take pride in their blackness while relying heavily on African values and culture as opposed to the European colonization. The French colonial racism is strongly withstood by these black people with their shared cultural heritage by forming a collective black identity. Thus the colour of the body that has previously been the ploy to draw the line of demarcation becomes an emblem of protest against racism. The assertion of own ethnic identity against the colonial superiors is also made possible by Benjamin Zephaniah, the British dub poet of Jamaican origin as he adopts the popular form of performance poetry in Caribbean Island to raise several pertaining questions against racial discrimination. In one of poems titled as *The Race Industry* he points out how racism thrives in a state that sanctions and perpetuates it as he writes---"We want more peace they want more police./ The Uncle Toms are getting paid./The race industry is a growth industry." (Zephaniah) That he is concerned with the institutional racism is quite evident in these words as he refers to the failure of the authority to stop discriminating among the people for their colour or ethnic origin. The fallacy of Enlightenment reason is alluded in this way to expose the pitfalls of liberal ideals of the Empire. Zephaniah who is influenced by Jamaican poetry and music becomes popular within African-Caribbean and Asian community for his poetry. He succinctly attempts to write what he himself strongly believes and this is why his poetry is invariably sharp in locating the root cause. Benjamin Zephaniah curtly brings attention to the structural racism in the poem as he writes, "We say sisters and brothers don't fear. /They will do anything for the Mayor./The coconuts have got the job./The race industry is a growth industry." Thus he harps on how racism is infused within the structure of the state politics that legitimizes the discrimination. A Rastafarian poet Zephaniah thus reacts vehemently against the racial profiling as state policing penetrates into racial identity. As an incorrigibly rebel poet he focuses on how black people are interpellated— "They're looking for victims and poets to rent. /They represent me without my consent." (Zephaniah) This dub poet moves to London from his home town Handsworth on a mission to reach to larger audience. Subsequently he becomes immensely popular for the dub music as his poetry is heard in public demonstrations against unemployment, homelessness and neo-fascist white supremacist National Front. Thus he achieves success in popularizing poetry in United Kingdom by instilling his ethnic essence in it. The fact that racism is inextricably linked with everyday existence is suggested by the radical poet who writes--- "In suits they dither in fear of anarchy./They take our sufferings and earn a salary./Steal our souls and make their documentaries./ Inform daily on our community./Without Black suffering they'd have no jobs./Without our dead they'd have no office./Without our tears they'd have no drink./If they stop sucking we could get justice./The coconuts are getting paid./Men, women and Brixton are being betrayed." (Zephaniah) In this way Zephaniah purports to convey that there is a coercion inflicted on the black people and this act of violence is legitimized by the colonial rule that normalizes the colonial violence by typecasting the black people in order to maintain the domination. However, there is no underlying suggestion made by the poet to resist against such categorization. He only lays bare the sheer injustice meted out to the blacks who are compelled to undergo excruciating pain throughout their life. This is perhaps why Benjamin Zephaniah who later becomes the poet laureate and is nominated as one of the most important fifty post-war poets of England rejects OBE from Queen Elizabeth II as he does not want to get recognition from the monarch while remembering the fact that his black forefathers and foremothers had to suffer insurmountable torture few generations back during the colonial rule. The victimization of the black people is still fresh in his memory and he rebels against it in his own way. The wound leaves an indelible mark on his body as he struggles against this process of marginalization by writing these dub poems which are ethnocentric in detail. The categorization of black experience under one unifying framework is itself an attempt to erase the heterogeneous ethnic identities and Zephaniah carries out his fight against this politics of representation in discursive practices.

The stereotypical black identity turns out to be the contested arena to figure out the loopholes of enlightenment logic in fetishizing the black imagery via normative principles. The culturally constructed categories of black identity are made visible by essentializing the notion of race and ethnicity. Zephaniah denies the fact that he consciously writes about politics while dealing with the suffering of black people. He does not want to involve himself into composing anything that becomes outwardly political and the experience of black people varies as Jean Stephancic and Richard Delgado writes, "...the idea that each race has its own origin and ever evolving—is the notion of intersectionality and anti-essentialism". (Stephancic and Delgado, n.p) What he intends to do instead is hinting at the black cultural politics against the colonial valorization of fixed racial categories in its hegemonic liberal discourse. The constitution of black cultural aesthetics is made even more explicitly visible in another poem by Benjamin Zephaniah titled as *We Refugees* where he writes, "I come from a musical place/Where they shoot me for my song/And my brother has been tortured/ By my brother in my land./ I come from a beautiful place/ where they hate my shade of skin/ They don't like the way I pray/ And they ban free poetry." Thus this direct narration unveils the extent of exploitation of the blacks as the poem situates the black experience within colonial confrontation. The recurrent corporeal imagery in this poem rearticulates the assertion of the black poet who falls back on his own tortured body to document the injustice. The black people are always under the surveillance of white colonial masters and in this regard the poet rewrites the body of the coloured people to resist the misrepresentation. The poet also raises his voice to allude to the repressive action of the state to encroach upon the creative activities of the black people as nation-state always strives to curb the voice of protest emanated by any cultural form. As an immensely successful poet in United Kingdom this is quite an incredible fact that Zephaniah achieves popularity even when he refers to the state control over the creative force. The poem becomes a form of protest that is occasioned by resorting to a popular mode of performance which has an immediate appeal to the white people. Benjamin Zephaniah does not only introspect into the racial indiscrimination while composing these dub poems but he is also concerned about several other issues like the representation of black women and the displacement of the native people as he writes--- "I come from a beautiful place/ Where girls cannot go to school/ There you are told what to believe/ And even young boys must grow bears./ I come from a great old forest/ I think it is now a field/ And the people I once knew/ Are not there now." (Zephaniah) Thus the lament over the displacement of the people creates an interrelation between the self and the place. History is embedded in the place as the spatial metaphor signifies the power exerted in controlling the colonized subject. The narrative of displacement recuperates the spatial politics as the process of history is inscribed within it. Thus the poem encompasses the broad span of history to portray the victimization of the native people who always suffers from a sense of not belonging to any place in a colonized country as the poet writes, "We can all be refugees/ We can all be told to go/ We can be hated by someone/ For being someone. / I come from a beautiful place/ Where the valley floods each year/ And each year the hurricane tells/ That we must keep moving on." (Zephaniah) The agony of the placeless blacks is perceivable in these lines as the place is constantly rewritten by the poet who aims to guard against the colonial control over place in Eurocentric discourse. Thus the narrative of alienation turns out to be a potent medium of protesting against the desecrating nature colonial invasion. The poet does not romanticize the past days but aligns himself with his cherished ethnic origin---"I come from an ancient place/ All my family were born there/ And I would like to go there/ But I really want to live." (Zephaniah) This is the perennial cry of the displaced subject as his life gradually becomes fractured and fragmented without any solace. But the disrupted existence of the blacks is articulated in this poetic form which recovers the dislocated people by acknowledging the afflictions they are made to suffer from. The recognition of the fact that the blacks are castigated as an heinous race that should be erased from history altogether is clearly evident in these lines---"I am told I have no country now/ I am told I am a lie/ I am told that modern history books/ May forget my name." (Zephaniah) The process of obliteration is a commonplace phenomenon in the history of mankind as the extermination of the Jews also comes as a corollary in this regard. The attempt to write off the coloured people by the West is countered by reaffirming the faith in struggling against any oppression as the poet utters, "Nobody's here without a struggle,/ And why should we live in fear/ Of the weather or the troubles/ We all come here from somewhere." (Zephaniah) In another interesting poem titled as *The British* Zephaniah imparts the cultural diversity of Britain that provides shelter to multiple ethnic people as a melting pot ranging from Angles, Saxons, Jutes, Normans through Chileans, Jamaicans, Dominicans, Trinidadians, Chinese, Ethiopians, Vietnamese to Indians, Srilankans, Nigerians, Pakistanis, Malaysians, Bangladeshis, Spanish, Iraqis, Afghans, Palestinians and such numerous such other. What is important here is the food imagery that is deployed by Benjamin Zephaniah who writes towards the end after arranging all these ingredients of ethnic identities---"All the ingredients are equally important. Treating one ingredient better than another will leave a bitter unpleasant taste" and rounds off the poem with a caveat---"An unequal spread of justice will damage the people and cause pain. Give justice and equality to all." Thus the co-existence of different cultures is only possible if there is justice blissfully meted out to each of these ethnic groups. Though this may be tinged with ethical excess it has always been the urge of the people with a fairly long colonial history.

Works Cited

Stephancic, Jean and Richard Delgado. *Critical Race Theory: An Introduction*, 2nd ed. United States: New York University, 2001. Web. 10.4.2015

Zephaniah, Benjamin. RHYMIN. <benjaminzephaniah.com > Web. 10.4.2015

Chapter 12

READING YEATS UNDER POST-COLONIAL LIGHTS

DEBALINA ROYCHOWDHURY
M.A. (Double), M.Phil. in English Literature
Senior Lecturer, English
In-charge of Entrepreneurship Dev. Cell
Convener, Placement Committee.
Elitte Institute of Engineering & Management

W.B. Yeats falls under the "Problematic and conflicting time of literature. His poems, I found to be most dynamic in thoughts. Yeats' poems incorporated the luxuriant flavor of Romanticism. The spirit of Victorian period, the realist discourse of modernism and the postcolonial perspective. The political turmoil of Ireland during the period of Yeats gave vent to the postcolonial overtones. some of his poems imbibes the tune of post colonialism. The subtle pain of suppression echoed in many of his poems.

In the word 'Post – Colonialism' the prefix 'post' suggests the paradoxical 'in – betweenness'. It is a combination of love for nation and freedom, pride of nationality and environment. Post colonialism stood opposed to colonialism both during and after the colonization. The fate of Ireland during W.B Yeats was not at all pleasant. It was under the British colonies. The historical Celtic was a big bang at that time. This was a movement for resuscitation of the heritage and, to bring about a unity of being. This gave birth to the national spirit and many of the works of this period carried the love and dedication for homeland. The contemporary litterateur and dramatists expressed their patriotic feeling for Ireland and Irish culture to save them from the English clutches. This nationalism or patriotic feeling against the British rule was nothing but postcolonial approach.

Yeats' nationalism was primarily literary and artistic more than political. 'Post Colonialism' according to the critics denotes the period after colonial period and that which is opposition to colonialism. But critics of recent times are towards the view that post colonialism also refers to that which belonged during colonial set up but expresses from the contrary point of view. The patriotism, nationalism of the period expressed in the colonial period is deliberately postcolonial in nature. Again that does not mean I am to prove Yeats as a postcolonial poet no he is one of them undoubtedly. This is just an analysis or study of his poems from post colonial perspective. Before going deeper into the aspect, it can be put up that Yeatsian poems portray the then social unrest and his nationalism. Thus for Yeats, it was more literary and artistic than being political. That does not suppress the political facts and happening of the contemporary period.

The most important poem in this context which shall be discussed foremost is Eater 1916. This poem has a deep historical value and is at tribute to the heroic martyrdom of the Irish nationalists and patriots. This poem was composed when the Irish patriots faced terrible odds and led by Patrick Pearse the movement resulted in their death. The rebellion of the Irish nationalists leading to success in saving Ireland from going under British Rule. This was known as the Easter Rising. The nationalists defied the British dominance and fought for the Irish republic. The nationalists were executed but the movement proved to be success. Yeats' poem subtly focuses the outcome as

All changed, changed utterly,
A terrible beauty is born.

He refers to all the martyr nationalists Pearse, Connolly, MacBride who contributed their best for the sovereignty of Ireland. He also says about other people too those who were intricately attached which the Easter Rising. References to women like Countess Markiwiez Maud Gonne who directly contributed to the movement. Yeats is more a poet than a nationalist.

Yeats Poems that death with the memories for those who sacrificed their lives the nation and their commitment is a subtle Post- Colonial feeling that came forth even during the chaotic period of colonization. The deepening disintegration that resulted massive for Ireland. The conviction rose out of this situation that created a vision of an upcoming antithetical civilization to put an end to the existing one. ---- The second Coming. This ending of one age at the close of other and the primary antithetical nature of it is described in different ways by different critics. I found it a transition Colonial to Post-Colonial social system. To stress this view I would quote from the poem "The Second Coming"–

The falcon cannot hear the falconer

Things fall apart; the centre cannot hold (ll – 2-3)

The 'falconer' obviously indicate the control, the power centralized and the 'falcon' is going away from the hold signifies the repulsion and surging away from the hold signifies the repulsion and surging away from the control unit; as if the subjugated the colonized moving away from rule of the colonizers. "The blood – demanded tide" in which "ceremony of innocence is drowned" The supporting lines that strike the keynote are surely some revelation is at hand surely the second Coming is at Land.

Yeats' nationalism and patriotic fervor gleams in his poem 'The Rose Tree'. Dramatically the poet present tow great leaders of Easter rising Pearse and Connolly conversing about national symbol of Ireland,. Rose Tree. The withered tree suggests the dying condition of Ireland and its national degradation. Then added to the dramatic essence James Connolly is made to speak about the nourishment of the tree suggesting the national health. The line that says of replenishing the state by their blood brings the prominent patriotism of the poet with a mild symphony of post colonial order.

W.B Yeats was not only a Modern poet or one of the Last Romantics. He belonged to the epoch of conflict, disintegration and stressful colonization; thus his poetry imbibed in them the patriotic ardency and a poetic revolt against colonization.

Yeatsian symbols are very interesting in his poems. The Leda symbol among other ones I found quite significant. It carries multidimensional explanations. Mythological character Leda, the beautiful princess who was raped by Zeus is a recurrent emblem in his poetry. The poem 'Leda and the Swan', portrays the morbid molestation of Leda which is interpreted differently by different critics. Leda can be read as a feminized and subjugated manifestation of Ireland and on the contrary, the disguised Swan becomes the representation of colonial Britain. The painful torment is picturised in such a subtle and morose way as if the poet felt the pain himself. Along with this, the line,

Did she put on his knowledge with his power
Before the indifferent beak could let her drop?

Leda is the symbol of impending devastation. The rape of her and the birth of Helen and Clytemnestra and consequently their contribution in the fall of Troy and the Agamemnon's death indicates destruction of humanity leading to a creation that begets annihilation of humanity in a larger scale. In the quoted lines it is hinted that birth of a new Ireland is possible soon after the withdrawal of England (dropping from the "indifferent beak"). It refers to the pre-colonial Ireland that after suppression is led to the Civil war.

Post colonialism asserts the cultural integrity. This kind of literature, seeks to determine the richness and validity of indigenous culture to restore pride of the nation that gradually degraded and faded under the colonial domination. In poems of W.B. Yeats, the promotion of Irish culture is prominently observed. The celebration of art, culture, popular legends of Ireland are found vibrant. The legend of Maeve, Cuchulian Legend, Fergus, Deidre, direct and indirect references of Irish nationalists as legendary heroes, remaking the Irish myths is seen as his postcolonial fervor. The characters like Oisin, Conchobar and Ribh figure out the Irish Culture against the dominance of British rule.Oisin, the Fenian hero, bears the Irish culture and thus propounded in Yeats' poem. Made immortally young by the fairy spell and even after visiting fairy ecstatic lands, still yearns for a return in the world of humans though old. Having lost the magic charm, bereaved of his eternal youth retreats in the human world down with age of three hundred years, old, decrepit after undergoing a sad romantic journey in three different lands and surging back tohis own land brings out the patriotic feeling and nationalism. Cuchulian Muirthemne, another legendary Irish hero, bright in his supernatural origin, gaiety, and isolation is a blend of the poet's personal and cultural characteristics. He is a proportionate combination of mysticism, nationalism, and heroism. Ribh another mystic character, who elaborates the philosophy of sex in relation to the orthodox Christian society. Ribh's idea of spirituality is more like that of Indian metaphysics. The significance of 'Atman'- the Self is heightened beyond all bounds. These fables portray the deep and exalted Irish culture that was to be restored to keep away the British culture and this was undoubtedly a step of a postcolonial writer.

Revising history is also one of the essential traits of Postcolonial literature. In response to the "outside history," its progress and authority on unchanging timeless societies, postcolonial works designs areas telling and retelling the history and legends of the colonized society or nation. The reference to the classical history and mythological adaptations found in the poems stand contrary to the British culture. The very little of the poem "The Municipal Gallery "Revisited" brings the faded sound of history. W.B Yeats gives an account of his personal feeling when he surges thirty years back. He visits the Art gallery in Ireland and sees the portraits of the museum. His personal grief mingles with the history- consciousness as the people for whom Yeats feels sad are of historical importance. He introduces, to the readers the great men and women who owed much to Irish history.

> Griffith staring in hysterical pride
> Kevin O'Higgins' countenance…
> (ll-4-5, Municipal Gallery Revisited)

Arthur Griffith the patriot, Kevin O'Higgins the great intellect and unpopular soldier to whom he pays his respect. He addresses the unidentified soldier as "revolutionary soldier." The picture of Higgins was painted by Lavery. Yeats refers to a woman's portrait of who is too a lady of historical importance. She is, according to Arland Ussher, lady Charles Beresford wife of William de la Poer. Yeats says that he met her 'fifty years ago'. Robert Gregory and Hugh Lane is also introduced. Major Robert Gregory was the son of Augusta Gregory who died in an air crash. Hugh Lane is another young man who is lady Gregory's nephew. Yeats also refers to lady Gregory who had played an important role in his life. Her portrait was painted by Mancini. He refers to her even in the fifth stanza as "….that woman in that household…". Yeats refers to the estate of lady Gregory at Coole Park where he had stayed and visited often. From the factual history he shifts to the fictional past; Yeats refers to Antaeus to symbolizes power and strength Antaeus is the son of Poseidon and Earth. According to the legend when he was attacked by Hercules, drew new strength from his mother whenever he touched the ground. This is also Yeats' patriotism and the word 'contact' specially with earth, is a concept of patriotism and love for Ireland. This can be considered as the postcolonial awareness he intends to spread.

The "Quattro cento finger" in the poem "Among the School Children" conjures up the specified symbol of classical beauty; it is another history conscious imagery of Yeats. The word is related to fifteenth century, Italian genre of and art. Yeats also involves great philosophers of past - Plato, Aristotle and Pythagoras. Plato was the great thinker who explained the world as a silhouette of ideas in more or less all the branches of studies, was also the tutor of Alexander and thrashed him to teach. These characters open up the golden leaves of history, the great kings and their greater tutors. He also refers to Pythagoras, the great musician, and philosopher and mathematician. It is admitted that Yeats accepts the predominance of death upon these great men. The continuous references of Iliad in various ways are suggestive of an impending war against the infiltrated power. 'Leda' becomes a reference mark for Yeats which bears a lot of explications, such as beauty and sexuality, productivity and source of destruction, symbol of oppressed nation. The Sphinx in "The Second Coming" significantly highlights the arrival of a terrible social change that would stand binary to the conventional social system. The historical allusions are continual in his poems delineating his disgust for the colonial suppression and anticipating a drastic change a counter to the suppression of humanity.

Yeats compares himself with 'a weather-worn marble triton' in "Men Improve with the Years". The word 'Triton' suddenly opens before us the verdurous world of Pagan mythology where mature was all-pervading, most potential. 'Triton' is a Greek deity holding a trumpet made of conch shell. He is known as the son of Poseidon, The sea God and Amphitrite. But, despite of the image, the whole expression rouses a range of overtones. If 'weather' is considered as the vicissitudes of life, or more specifically time and age—then the 'Triton' here, is old, weary and inert. 'Marble' again, suggests inactivity, coldness and lifelessness. 'Triton' in Greek myths are drawn as a merman whose upper portion of the body is like a man and the lower part like fish. The figure and the whole expression suggests a trenchant imagery of sexual incapability. To delve more deep in the imagery, this 'unhappy inertness' also bears the unwilling acceptance of the colonized nation- the power, the domination of the other. 'Weather' again suggests the adversities occurred due to the colonization. The 'Triton' holding a conch is suggestive of a dormant power and not a dead one. The 'trumpet' becomes a device initiating war may be in posterity by a clarion call from it. Thus, the image of a 'once potential'

(Triton) now transformed in marble statue can be substituted as a colonized race but having the power to withstand though latent now.

Post colonialism can be termed as a quest for one's origin, the root that signifies a culture, a legacy that is suppressed under the prey of colonizers. With Yeats it becomes subtle and profound. The poet designs his poems in such a prismatic manner that multifarious implications like hues manifold reveal. I would prefer to conclude with a vividly eloquent line of Yeats, that strikes a fathomless strain of human existence, unperturbed acceptance of reality and his patriotism for Ireland. The 'horsemen' is a characteristic epitaph that symbolizes Irishmen in general. According to T. R. Henn this suggests respect and awe for his own land as these lines, Yeats instructed to inscribe on his grave as an epitaph:

Cast a cold eye
On life, on death.
Horseman, pass by!

Work Cited:

Yeats, William Butler, Edited by Watts, Cedric: The Collected Poems of W.B. Yeats, Wordsworth Poetry Library, Hertfordshire, 1994

Albright, Daniel: The Myth Against Myth, Oxford University Press, 1972

Rai, Vikramaditya: Poetry of W.B. Yeats, Doaba Publishing House, 1971

Wright, George T. The Poet in the Poem; Perspectives in Criticism. University of California Press: California, 1960

Printed in the United States
By Bookmasters